A Book of Migrations

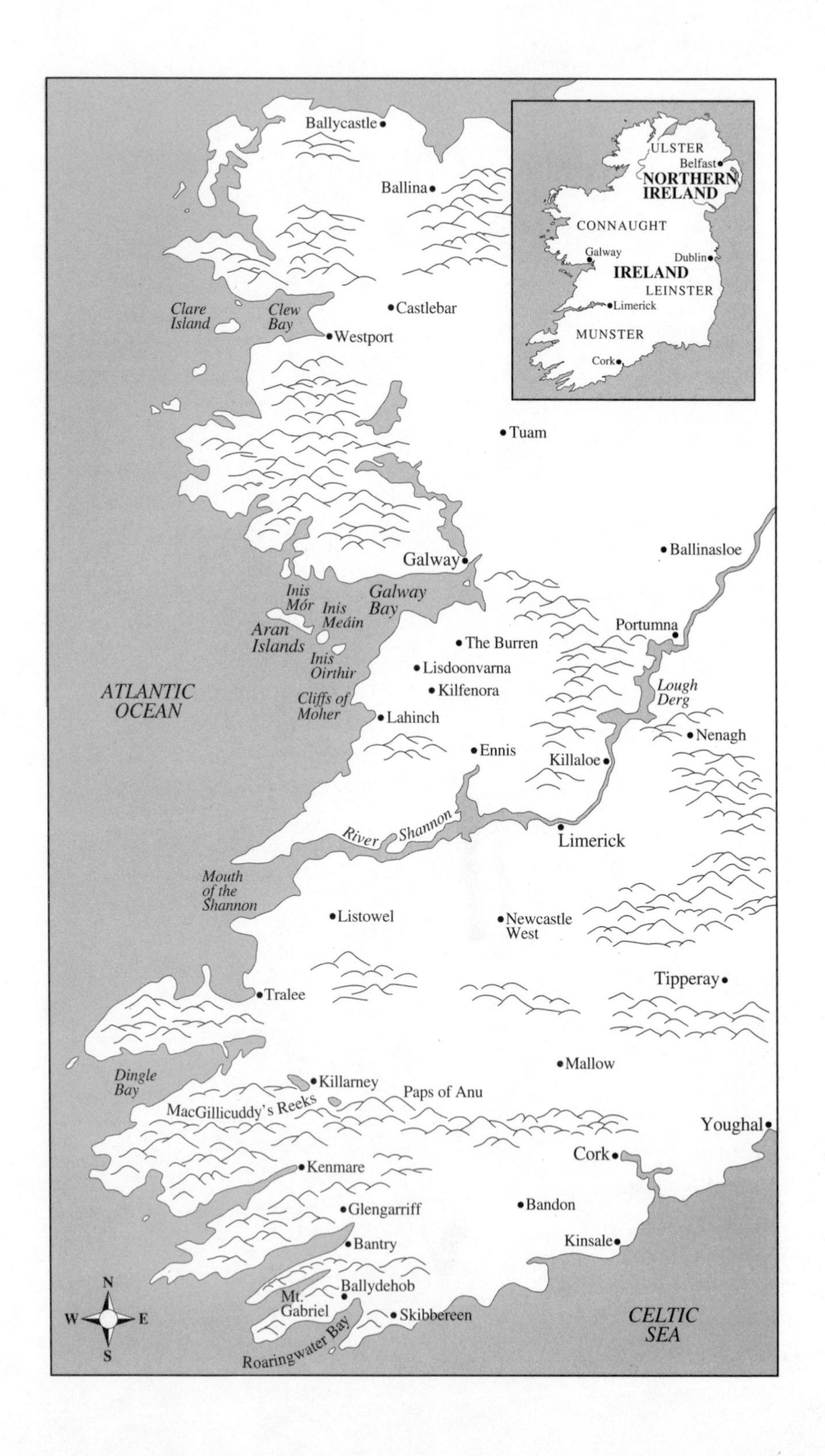

Ballycastle
Ballina
Clare Island
Clew Bay
Castlebar
Westport
Tuam
Galway
Ballinasloe
Inis Mór
Inis Meáin
Galway Bay
Aran Islands
Inis Oirthir
Portumna
The Burren
Lisdoonvarna
Kilfenora
Lough Derg
ATLANTIC OCEAN
Cliffs of Moher
Lahinch
Nenagh
Ennis
Killaloe
River Shannon
Limerick
Mouth of the Shannon
Listowel
Newcastle West
Tipperay
Tralee
Mallow
Dingle Bay
Killarney
Paps of Anu
MacGillicuddy's Reeks
Youghal
Cork
Kenmare
Glengarriff
Bandon
Bantry
Kinsale
Ballydehob
Mt. Gabriel
Skibbereen
CELTIC SEA
Roaringwater Bay
N
W
E
S
ULSTER
Belfast
NORTHERN IRELAND
CONNAUGHT
Galway
Dublin
IRELAND
LEINSTER
Limerick
MUNSTER
Cork

A Book of Migrations

Some Passages in Ireland

REBECCA SOLNIT

London • New York

First published by Verso 1997

Verso
UK: 6 Meard Street, London W1V 3HR
USA: 180 Varick Street, New York NY 10014–4606

Verso is the imprint of New Left Books

ISBN 1–85984–885–0

British Library Cataloguing in Publication Data
A catalogue record for this book is available from the British Library

Library of Congress Cataloging-in-Publication Data
Solnit, Rebecca.
A book of migrations: some passages in Ireland/Rebecca Solnit.
p. cm.
ISBN 1–85984–885–0 (hb)
1. Ireland—Description and travel. 2. Solnit, Rebecca—Journeys—Ireland.
3. Literary landmarks—Ireland. 4. Historic sites—Ireland. 5. Walking—Ireland.
I. Title.
DA978.2.S63 1997
941.5—dc21 96–54867
CIP

Map by Bart Wright
Typeset by M Rules
Printed by Biddles Ltd,
Guildford and Kings Lynn

Contents

Acknowledgments

This book goes out with thanks to the people who laid its foundations – my uncle, Thomas Davis Allen, whose identification with Ireland and genealogical research there provided me with the cypher of Irish citizenship; my mother, Theresa Allen, who was, during my childhood, working for civil rights and urging me to draw oak trees; and Lee Snodgrass and Paddy O'Leary, whose friendship and storytelling verve on my first visit to Ireland were among the strongest inducements to return – as well as to Verso's American editors, Mike Davis, who was encouraging about this book when it was nothing more than ideas and notes, and Michael Sprinker who was there throughout the writing. I also thank Sister Kathleen and the Portumna Sisters of Mercy and Cathleen MacDonagh and her family for their hospitality on my journey; Ray Ryan of Cambridge University Press who read the manuscript and made valuable suggestions; Bonnie Nadell and Irene Moore of Frederick Hill Associates Literary Agency; Alice O'Malley; Lucy Lippard, for whom I housesat while writing the first chapters in late 1994 and some others in the summer of '95; Bill Studebaker and Brenda Larsen; Pame Kingfisher and Kiya Heartwood who, in an early version of my travel plans, were to accompany me and who had much to say about ethnicity, music, and memory; and various others whose conversation and friendship have been important: Lewis deSoto, Tim and John O'Toole, Sono Osato, Sarah Wright, Dana Schuerholz, Dianne Driscoll Nepali, Valerie Soe, my brothers Stephen and David who were, respectively, organizing for immigrants' rights and the rights of the homeless as I wrote, and, last but most, the lovely Pat, snake charmer and hillwalker, formerly of Joshua Tree. Finally, I have been meaning since my first book to credit the city of San Francisco's rent control policy, which has helped secure for me the increasingly rare circumstances in which independent research and writing can be done.

. . . the past is a country from which we have all emigrated . . .
Salman Rushdie

Every day is a journey, and the journey itself is home.
Basho

Preface

On the last day of 1986, I became an Irish citizen. My newfound status as a European has not yet ceased to bemuse me – my purple passport with its golden harp seems less like a birthright than a slim book on the mythologies of blood, heritage, and emigration. And so a few summers ago I went to this foreign country that claims me to think about these things. This is not a book about Ireland so much as it is a book about a journey through Ireland. I hasten to disclaim any great authority on the subjects of Irish history and culture, though my literal, geographical journey was overtaken by two longer journeys: through readings in Irish history and literature and through the writing itself (to say nothing of a follow-up trip to round out my researches). This is likewise no core sample of contemporary Ireland; in the same spirit Irish tourists may head straight for Graceland, I took off for the places that appealed to me and let attraction and invitations stitch together the rest of my route.

Ireland delighted me by offering so many stories and circumstances in which individuals and populations were fluid rather than ossified, undermining the usual travelers' dichotomy of a mobile figure in an immobile landscape. It was this play between memory, identity, movement, and landscape that I wanted to explore, and the ebb and flow of populations that constitutes invasion, exile, colonization, emigration, tourism, and nomadism. In other terms, this excursion offered me a whole set of linked possibilities: to muse about the identity politics raised by my American activism and my Irish passport; to rethink my education in English literature – to separate the status of great literary works from that of great landmarks of civilization, for which Ireland as a kind of backstage to the drama of imperial Britain is ideal; to look at a place where the tidy phrases of contemporary conversation – terms like *native* and *white* and *Europe* and *first world* – begin to fray and come apart.

Travel is also a psychic experiment. In different places, different thoughts emerge, and this too I wanted to trace. I tried to use the subjective and personal not to glorify my mundane autobiography but as a case study in how one can explore the remoter reaches of the psyche by wandering across literal terrain. This book is itself not a travel book in the usual sense, but a book of essays sequenced and shaped by my journey. That journey is the continuous thread running through the chapters, which can be regarded as variously faceted beads, each one made differently from the raw materials of landscape and identity, remembering and forgetting, the fixed and the fluctuant, and out of travel itself – out of my own modest passage and all the great tides any such journey echoes.

1 The Cave

She grasped the windowsill and leaned across me to look down, and the veins on the backs of her hands looked like earthworms, the ones that on the rainy days of my childhood swam to the sidewalks and lay there exhausted and rosy gray. Thirty thousand feet below, beneath torn veils of cloud or fog, a wild landscape full of pale blue lakes stretched, some portion of what we surmised was northeastern Canada. Gazing out of the windows of airplanes, I always try to memorize the shapes of the bodies of water that lie below, but if I ever get to an atlas at all, the strangely alluring forms I saw never resemble the neat, familiar world of maps.

The woman with the veiny hands talked to me on and off all the way from San Francisco to London, and her even more weatherbeaten husband with his crewcut and trembling hands never spoke at all. He'd been silenced by a recent stroke and needed help getting up and manipulating all the paraphernalia of airline journeys, the seatbelts and headphones and trays, so I was boxed in by them for twelve hours. He had settled into his inarticulateness like a dog among people and watched us as elements in a life that no longer included him. She had that garrulous sweetness so many women of her generation do, as though they were still waiting for life to begin, fresh, eager, and sheltered. They were going to London to join a tour of England; I was going to catch another plane to Dublin as soon as ours landed and spend my time in Ireland. Snatches of a movie about innocent Irish Catholics framed, tortured, and imprisoned by the British government not very long ago drifted into our conversation from the tiny televisions mounted in the backs of the seats, an odd choice for a flight to London, more warning than invitation.

She wanted to know if I had a dream to fulfill in Ireland. I thought of all the complicated things I could tell her and said I was going to walk across its west. By yourself, she exclaimed, and By myself, I replied, and turned the conversation back on her. She told me that if I ever came across any churches of the Latter Day Saints on my pilgrimage, I should go in and have another kind of adventure, about as polite a gesture of evangelism as I've ever encountered. When she let on they were Mormons, things began to fall into place. Although she had first said they were from Berkeley, it turned out they were originally from the Great Basin, as their faces and manners suggested. I wanted to hear her talk about it; I like to hear people talk about certain places the way some people like to hear their beloved mentioned; even the names of secondary highways and small towns can summon back the place itself. She added that there was no place on earth so beautiful as Logan, Utah, in the spring, although that was perhaps because she was twenty then, and I could guess at the lilacs and frame houses and wide streets and wider sky. I like Mormons too, as the only immigrants to love the northern deserts of the United States and to truly settle the West at all. They did it by fantasizing the place as the new promised land and themselves as the new Israelites at the end of their exodus, so an Old Testament, old world veneer lay over the Utah they settled in 1846 and thereafter – but they stayed and made a home of it, which is more than one can say of most white people in that West.

This old woman traveling to green England and yearning for arid Utah seemed, when I thought back later, to be where my journey began. Or it began earlier that spring near Twin Falls, Idaho, not far from Logan, Utah, in the kind of sagebrush prairie that I pictured when she spoke to me of Logan. When the philosophy professor at the College of Southern Idaho had called me up to ask me to come to Twin Falls and talk about landscape, she was audibly surprised that I had been there before and would be happy to return. The Snake River's Shoshone Falls, right outside town, is one of the country's more spectacular waterfalls. The black volcanic canyon of the Snake forms the northern boundary of the Great Basin, though the same terrain of huge, drab expanses, bathed in the incomparably clear and constantly shifting light of the arid West, stretches further north. Had the falls been somewhere else, they might be a major tourist attraction, another Niagara, but the Snake is used instead as a giant irrigation ditch, and when I'd passed through the summer before, most of the water had been diverted to crops, and the black rocks beneath the whitewater jutted out like beggar's bones. When I returned, there was still snow on the ground, and coyotes frolicked with each other on the edge of the local airstrip.

I stayed an extra day, and Bill Studebaker, the honors program chair, and Brenda Larsen, the philosophy teacher, took me out to see the sights. Had the dry air and easygoing voices not told me I was far from the urban centers of the coasts, the fact that these two professors had time and hospitality enough to spend a whole day driving a stranger around would have been enough. Brenda and I had a conversation to resume: she wanted to compile a collection of philosophical paradoxes, and I saw in it some affinity to my concern with metaphors. The day before, as we picked our way among the iced-up pools along the banks of the Snake, we had begun to approach some definitions that made us happy, and we refined them over a tomato curry she cooked me that night, this pretty woman whose appetite was equally sharp for food and ideas. Our metaphors and paradoxes seemed like ways of being in two places at once, and we had begun to suspect that the philosophical desire to reach some final destination is, finally, a humorless attempt to be in only one place, the unreachable vanishing point of literal truth itself. Arrival, like origin, is a mythical place.

Brenda and Bill came and got me early in the morning, and we went eastward past the farm fields, through the town of Eden with its inexplicable First Segregationist Fire Department, to the open rangeland. A local joke had been trotted out for my benefit the night before, about Twin Falls being situated halfway between Bliss and Eden; and Bill had published a book of poetry called *North of Bliss.* He was weatherbeaten with a crewcut, a younger version of the man on the plane, and he had the wary quality of creatures who aren't sure they want to come in from the wild. He had a wife too, a Mormon Shoshone woman from the nearby Fort Hall Reservation, and four children, but he still seemed withdrawn and private somehow, one of those autodidacts who will tell stories, suddenly and unpredictably, but don't exactly converse. That day, out of the blue, he said, 'One Hunter and Jaguar Deer,' the names of the twins who defeat the old gods of Xibalba in the Mayan creation epic, the *Popol Vuh,* adding that he'd like to be named One Hunter. Another book of his had been on the petroglyphs of the region, which is the northern end of the vast Shoshone homeland, and, as we drove down long straight numbered roads and snaking dirt roads on the high bowed plateau, sometimes slanting downward into farm crops and sometimes upward to crests from which the distant ranges that framed the horizon could be seen, he talked about it.

When the potholes got too deep for the lowslung university car, we left it in the middle of the dirt road – no one else was likely to come by that day, maybe that week – and walked in the cool clean air, with old snow and clumps of still-gold grass and black lava outcroppings scattered across the land on either side. We saw a cow pie in the road with a coyote turd on top of it, a territorial marker like a checkmate in chess, and a golden eagle sitting atop one of the volcanic piles not far away in this treeless landscape. Eventually Bill spotted the cave, which looked like just another lava outcropping on this arid prairie. Its rough roof tufted with scrubby plants provided a long view in all directions, and the little hummocks of rock that made it up fit together like the cells of a honeycomb or rolls in a pan. A depression like a streambed led down to a small opening, covered by a cyclone fence that had been burst open. Inside, the cave was a single hemispherical lava bubble, an ancient fiery gasp of the earth turned to a perfect stone dome perhaps forty feet across and nearly twenty at its highest point, dimly lit by the daylight. The ceiling was rough, but lacked those deep recesses, irregularities, and extrusions that make most caves so unnervingly mysterious, with passages that could harbor anything. Most caves feel awkward and alien and perilous, but this one – I could imagine living in it. In fact I found myself picturing how I'd arrange my furniture in it, with a fire venting at the mouth and the magnificent landscape outside as the new center of a starker, more serene world.

It had had occupants, Bill said, for the first seven of the last twelve thousand years. Then the region dried up, leaving the cave too far from any source of water to be habitable. The floor was soft dirt, the sifted result of excavations, and there were plastic tarps under it, their edges sticking out, so that if the excavations ever resumed, the archeologists could pick up where they left off. Nothing aboveground suggested the people who had lived here, no visible structures, no tombs, walls, carvings, no paintings in the cave, but it had been inhabited thousands of years before human beings are thought to have come to Ireland. This part of the world is old in ways that Europe is not, governed by a different sense of time altogether. But the traces of age are far subtler in most parts of North America, and newcomers easily imagine that it is new itself. Bill spoke of how the Shoshone traditionally destroyed the property of those who died, so that there were almost no heirlooms, no objects passed down the generations, no accumulations of property. I knew a woman in Death Valley, at the other end of Shoshone territory, who remembered such a funeral pyre herself, in the 1940s, when even the horse was killed. It was a circumstance that suited well the nomadism of the Shoshone in their arid homeland. Whatever its religious purposes, the custom kept people from piling up wealth over the generations, forestalling inheritance and the inequities that result. Nomads travel light on many levels.

The innumerable objects of the past are milestones telling us how far we have come on the one-way road of history. This destruction of property when it became the property of the past must have kept its practitioners living in a perpetual present, where all that existed of the past was what could be remembered and told. In such a scheme, culture itself was what could be carried in the head, and material culture was always being created anew from the surrounding landscape. In the oral history at the center of such a scheme of perpetual re-creation, the time between the dreamtime of creation and the present is immeasurable, elastic. The present is a mesa surrounded by creation rather than a slope rising up incrementally from the historic past. Forgetting, like remembering, has its uses, and the two are balanced differently in different places.

Bill had got into trouble, he said, for his conclusions in his book about the prehistory of the region. He had found that the region's art – the petroglyphs and stone tools – changed from a relatively refined state to the simpler work identified with the present-day Shoshone, and it seemed to him that they were relatively recent arrivals in the area. His conclusions were at odds with their sense that they had been here forever, since creation. For a lot of Native Americans, the Bering Land Bridge theory, which postulates that their ancestors came over from Asia and, over the millennia, migrated all the way to Tierra del Fuego, is as problematic as evolution is for fundamentalist Christians. Their religions often dictate that they were created where they are, with an indissoluble, irreplaceable symbiosis with their landscape. Once, when I was working as a Western Shoshone land rights activist, I stopped in at a gathering, not far from where we were wandering around the lava cave. There a Shoshone guy objected to both the Bering Land Bridge theory and evolution, saying with fine disregard for the official details, that Europeans might well be descended from Neanderthals – certainly it would explain why they were so hairy – but it had nothing to do with him and his people, who had never been in the old world, never been anywhere at all but the perfect dry bowl of the Great Basin, where they had been placed by the creator to take care of it and noplace else.

Politically and spiritually, such an approach declares that the relationship is untransferable, that they did not adapt but originated as they are, and therefore that there can be no replacement, no new home. There is a sense in which both versions can be true: the evidence of ancient migrations means one thing, a material history; and the relation of the people to their land means another, a cultural history. As long as their identity is so profoundly situated in their landscape, it is impossible to say that they were that people before they were in that landscape. If it is the genesis and face of their identity, it is literally where they are from, where they were created – which makes the rest of us rootless by comparison.

Even recognizing the existence of indigenous people in the Americas raises a lot of difficult questions about belonging for those Americans who descend from historical emigrations: questions about what it means to be and whether it's possible to become native; about what kind of a relationship to a landscape and what kind of rootedness it might entail; and about what we can lay claim to at all as the ground of our identity if we are only visitors, travelers, invaders in someone else's homeland. Some of the most literal-minded Euro-Americans have decided simply to become Native Americans, as though identity and heritage could be picked up in a process as simple as shopping, and the very shallowness of such methods of impersonation undermines the sense that there is something deep at stake.

A lot of Native Americans respond that Euro-Americans have their own cultural traditions, which they should recover rather than appropriating someone else's, and this too seems to simplify the matter. Many of us are children of refugees from countries that no longer exist, from atrocities no one spoke of, from traditions that had been trampled over earlier by the same forces; and in living on a new continent, most of us have begun to be something else, transplanted and hybridized. This evolving something else has never been resolved adequately, and perhaps it is irresolvable – unless *resolution* itself returns to its linguistic roots, which meant to unloose or dissolve, to clarify by liquefying, not solidifying. If being local is a matter of forgetting what came before, then the journey is completed by severing previous ties. *Naturalization*, the term for becoming a native, suggests this process of adaptation – and perhaps suggests that forgetting rather than remembering is central to an identity resolved like rain from a passing storm sinking into the soil.

I am in fact legally a European, naturalized, if not naturally. I have an Irish passport, thanks to some fancy detective work by my Uncle Dave, my mother's brother, who dug up the whole chain of birth and marriage certificates linking us to Ireland. That so thin a mythological fluid as blood entitled me to so solid a legal status as citizenship still amazes me. I am, I suppose, a third-generation Irish-American, and that old country is just beyond memory, with our paucity of family stories. I'm not much of an Irishwoman, let alone a Catholic, since my father's parents were immigrant Russian Jews, and I've been in hybrid California, world capital of amnesia, nearly all my life. My Irish passport seemed less like a legacy than a windfall, a key not to my own house but to an unfamiliar building, a return ticket not for my own journey but for that of my four Irish immigrant great-grandparents about whom I know almost nothing. According to the way people tend to talk about blood and roots and other charged images now, this unknown country is what's mine.

It isn't, but I thought that Ireland was a good place to think about it all. Ireland, which once formed the western edge of the known world and the wild west of Europe, which for all the teadrinking whiteness of its populace was colonized by England in much the same way as North America, which was the first of so many conquered countries from Algeria to Zimbabwe to liberate itself in the twentieth century, which is still described as the third world of northern Europe with its underdevelopment and huge unemployment, with its conflicts subsuming so many versions of cultural strife into a smaller compass of space and race, Ireland where Europe came closest to other parts of the world. Rather as the sweet-voiced Mormon was going to tour England with Logan, Utah, hovering somewhere in her imagination, I was going to wander Ireland and see how far away Ireland was, or how near. This other land in the guise of a motherland presented an opportunity for me to go both back and away, be lost and found, but mainly to think.

I wanted to think in a different landscape about questions that had arisen for me in my own: about the concentric circles of identity formed by memory, the body, the family; by the community, tribe or ethnic group; by locale, nationality, language and literature – and about the wild tides that have washed and wash over those neat circles, tides of invasion, colonization, emigration, exile, nomadism, and tourism. Ireland has been particularly shaped by these tides, and for me it seemed an ideal place to look at these things and their consequences. I wasn't sure what I would find – certainly I didn't expect to find a strong, if embattled, contemporary nomadic culture as I did at the end of my Irish expedition – but travel provides not confirmations, but surprises.

In a sense, it was travel itself that I was after. There's a convenient fiction preserved in travel literature, if nowhere else, that a person is wholly in one place at one time. The idea of nativeness is similarly a myth of singularity or perhaps an ideal of it, but even most contemporary more-or-less indigenous people have mixed ancestry, have undergone sudden and violent relocations, have lost something of their past in the process and picked up a lot from the dominant culture of the US, which is itself a hybrid with many sources, not all of them European. We are often in two places at once. In fact, we are usually in at least two places, and occasionally the contrast is evident. I always seem to be trailing through three or four at once: talking about Utah and thinking about Idaho caves while watching a movie about British prisons in an airplane over what was probably Newfoundland on the way to Ireland was nothing exceptional.

Once on a long trip down a wild river, I dreamed about my city and my home every night, and upon my return, I began to dream about the river over and over again. Here, most often, is nothing more than the best perspective from which to contemplate there: one climbs the mountain to see the valley. Traveling, I had found in the course of a year of farreaching excursions, shifted one's memory and imagination as well as one's body. The new and unknown places called forth strange, oft-forgotten correspondences and desires in the mind, so that the motion of travel takes place as much in the psyche as anywhere else. Travel offers the opportunity to find out who else one is, in that collapse of identity into geography I want to trace.

2 *The Book of Invasions*

The first thing I did in Ireland was miss the cousin who had come to meet me at the airport, one of the dozens of third, fourth, and umpteenth cousins my uncle had located on his genealogical spree a few years ago. This one was about to emigrate to New York to join the huge underground economy of undocumented Irish workers, and I was curious to talk to someone who was going to carry on the adventure of exile so long after my branch of the family had. So there he was, probably holding up a sign outside the baggage claim for me, and there I was, without anything more than a knapsack I'd carried with me, rushing onward toward Dublin and toward an odyssey in which I'd never track down any of the cousins I had scattered over the island, people with whom I shared nothing more than a sixteenth or thirty-second of a gene pool and a common past beyond our recollection. I got off near the Customs House in downtown Dublin, at a place where dozens of buses were belching their diesel breath into the dirty air, a stranger among strangers under a thin gray sky, with a tourist map to get my bearings.

Being in Dublin isn't quite the same as being in Ireland, though if it's an anomaly, it's one that includes at least a quarter of the three and a half million people in the Irish Republic. The city and the country are profoundly different everywhere, but in Ireland, the city is truly singular: a city, Dublin. Ireland overall seems to be a place where the industrial revolution has yet to happen, where the serene fatalism of rural time seems to govern the pace of life, where the wet green landscape is hardly ever out of sight and the tourist's picturesqueness is the native's isolation. My Irish history book from college said that the Celts' "way of life did not include the concept of towns"; the other population centers in the Republic – Limerick, Galway, Cork – feel like sprawling towns at best even now; and more than 40 percent of the population is still considered rural. But Dublin is a city, crowded and bustling.

It has been a city of invaders from its founding to its present, another reason it seems anomalous. Invasion has been the principal motif in Irish history, and Dublin the principal port of entry and site of occupation. It's a history that repeats interminable variations on invasion, its extended and usually brutal consequences, resistance, revolt, and flight – and contemporary Ireland has a curious blankness, as though it has outlived the only kind of political history it knows, while the conflict in the north reiterates the past on a reduced scale. The pattern seems set even before the historical invaders came, for the mythological history of the Irish Celts is the *Book of Invasions*, which recounts six legendary waves of incursions. This imagined Ireland is a place not of indigenous but of settled and assimilated immigrants and of older tribes whose faint traces underlie the present: the giant Fomorians, the small, cunning Fir Bolg identified with the precursors of the Celts, the beautiful Tuatha Dé Danann, children of the goddess Danu who dwindled over time to the stature of fairies, and the last to come, the Milesians. Of the Fir Bolg, says the *Book of Invasions*, "As everyone does, they partitioned Ireland," and of the Tuatha Dé Danann, "On Monday in the beginning of the month of May, to be exact, they took Ireland."

According to the archeological evidence, Ireland was first inhabited about eight thousand years ago, by the people who eventually built the magnificent stone circles and cairns that punctuate the countryside; the Celts who did not have the concept of towns invaded about twenty-four centuries ago. The Celts were, for the most part, left alone for their first Irish millennium. Although a Roman fort was very recently discovered twenty miles north of Dublin, the Roman invasion that must have produced it left no obvious traces in Irish history or culture. The imperial Roman vision of order that would shape and subjugate much of the rest of Europe didn't stick until the English brought it long after England's Roman phase. Christianity had come benignly, with St Patrick and various others in the fifth century, and when the rest of Europe was in its dark age of invasion and decay, Ireland was in its great Golden Age of wealthy monasteries and erudite monks who kept alive the European learning that was dying out out elsewhere (much modern Irish decoration on objects ranging from paper currency to manhole covers draws from the Celtic interlace of illuminated manuscripts and other art of this period, a sobering reminder of how long ago Ireland's last era of happy independence was). The significant invasions in that era were those the Irish carried out in Britain, most notably that carried out by the Scots, an Irish tribe that colonized the Pictish place now known as Scotland (and perhaps the raid on England in which the future St Patrick was captured and brought back as a slave). Ireland was a tribal society with a fine literature, art and scholarly tradition, and without a central authority, when the invasions began.

They began in the ninth century, when the Norse Vikings founded Dublin in a marshy region where the Pobble meets the Liffey. These Vikings kept the Celtic name of the place, Linn Dubh, or Black Pool, after the place where the two rivers met, and Dublin provided them with a base from which to loot and terrorize the rest of the island. The name outlasted them, for they were driven out at the beginning of the tenth century, only to be succeeded a few decades later by Danish Vikings, who stayed to develop a more extensive city. That city briefly reappeared when foundations were dug on Wood Quay for a new city hall in 1978, and was buried again under the international-style steel-and-glass offices, after wrangling over the fate of the site and hurried excavations. This lost city on the south bank of the Liffey had massive walls, but the houses themselves, woven out of wattles and plastered with mud, were hardly more substantial than basketwork. Viking Dublin, like dozens of later versions, lies dead and buried under its descendant, the modern city, whose physiognomy bears traces of its ancestors – a certain ancient building jutting out here, a street name and the angle of the street itself there. The Vikings had become craftspeople and traders and were assimilating into the local culture by the time of the next invasion, the one that determined Irish history until the twentieth century.

There's a poem of Jonathan Swift's, meditating on the drying up of St Patrick's Well in Dublin in 1729, a poem in which he implies that his native land was an Eden and pins its fall on England's invasion:

> *Britain*, by thee we fell, ungrateful Isle!
> Not by thy Valour, but by superior Guile:
> *Britain*, with shame confess, this Land of mine
> First taught thee human Knowledge and divine . . .

It is true that Irish missionaries had first converted England to their mild form of Christianity, and equally true that Ireland's peace and prosperity came to a conclusive end with the Norman invasion in 1170, only slightly more than a century after they had conquered England. And like the original fall from grace, Ireland's has often been blamed on a woman. In 1152, Devorgilla left her princely husband for – or was abducted by – Diarmait, the man who had usurped the kingship of Leinster, Ireland's eastern quadrant around Dublin. Not much is said about her, generally, except that she was forty-four at the time, her husband sixty, and her lover forty-two. So, in a middle-aged operetta that became a seven-century-long national tragedy, her husband eventually succeeded in driving her lover out of Ireland, and the despotic lover, Diarmait, sought help from the English king.

Minor invasions preceded the 1170 conquest, precipitated by Diarmait's promise to make a Norman lord his heir. A few hundred Norman-English knights, mounted, armored, and heavily armed, with about a thousand foot-soldiers, took Leinster for themselves in 1170, and made Dublin their headquarters. They built Dublin Castle, the great stone hulk that still broods over central Dublin, on whose walls the severed heads of rebels were once displayed, and other strongholds, parcelling out much of the land to Norman nobles. Their hold over the rest of Ireland waxed and waned while

Dublin remained essentially an English holding until the 1920s. By the fifteenth century, their grip on the rest of the country had slipped, and they held only the Pale, an expanse spreading about thirty miles around Dublin. Everything else was literally beyond the pale, a boundary that became a permanent fixture of the English language as the line between barbarianism and civilization, though the civility of the Normans was dubious at best.

Besides Dublin Castle and the two cathedrals, little can be seen of Dublin's past before the eighteenth century, when the present city was envisioned and built. There were recurrent invasions, insurrections, and bloodbaths, from the failed sack of Dublin by the Scots in 1317 to the uprising of the aristocratic and doomed Silken Thomas there in 1534. It was Cromwell's base for his mid-seventeenth-century anti-Catholic campaign. Order and a measure of peace returned with Charles II's viceroy in 1662, the Duke of Ormond. The expansive vision of this Anglo-Irish duke ushered in modern Dublin, with the creation of vast Phoenix Park and the first fine buildings whose pale façades give contemporary Dublin its neoclassical complexion. In the following century, the city belonged unequivocally to the Protestant businessmen and aristocratic administrators whose houses, squares, and Palladian civic buildings make it the finest Georgian city in Europe.

Their era came to an end with the century: in 1801, in the aftermath of increasingly viable rebellions, the Act of Union shut down the Irish parliament and merged its government with England's, so that Dublin was merely the principal city of a poor colony until it became the capital of the Free State in 1922. Almost all of Irish history is about its relationship to the larger island that on maps looks like a dragon looming over a lamb. The relationship is often analogized to that of a brutal man and a sadly oppressed woman. In his magisterial history of the potato, which is in large part also the history of Ireland, the English scholar Redcliffe Salaman describes the relationship between the two countries as an unhappy marriage, with the smaller island as the downtrodden wife. Of the Act of Union in 1801, he declares: "It was at best a *mariage de convenance* in which absence of affection, disparity of age, and inequality of fortune, were for the time overshadowed by the fear that if England did not secure her as a bride, her coquetry might end in bringing her paramour, France, unpleasantly near. . . ." Seamus Heaney describes it as a rape: "Ralegh has backed the maid to a tree/As Ireland is backed to England." It's a metaphor that accords well with an occupation that began with a deserting wife and ended among images of Ireland as Dark Rosaleen, Kathleen Ni Houlihan, and other romantic females.

Present-day Dublin lies on the map as though a drunk had tried to draw a grid using the Liffey as its horizontal axis, given up on geometry, and turned the grid into a spiderweb. The streets of the city center wobble along inside the ring of circular roads and canals, parallel streets warping into convergence. Few of them provide the long straight axes of Parisian or American avenues, so the Wicklow Mountains remain largely unseen around the city's western edges. Modern gridded cities put real estate first, with pedestrians consigned to the margins between repetitive city blocks, and compensate with vistas, but older, unplanned cities seem to have been laid out by walkers wandering off on centuries of private missions, with the buildings fit between their footpaths, a labyrinth of time and stone. Central Dublin itself is such a cobweb of elderly streets, which the statues stud like flies, and perhaps the streets were a flytrap, for most of the statues are of martyrs to their vision of Ireland, visions and martyrdoms often carried out somewhere within the city itself. To see the great population of statues one could almost believe that the citizens of this city turn to bronze after death, the way bodies tossed into Irish bogs turn to brown leather.

Though the statues mark history, they mark the present version. Earlier versions have been destroyed as part of Dublin's reinvention as a national capital rather than an invader's stronghold or a provincial headquarters: a statue of King William III, the Protestant zealot, was blown up in 1929, and one of George II on St Stephen's Green was destroyed to coincide with the coronation of George VI in 1937. The massive Nelson's Column on O'Connell Street was bombed in 1966, on the fiftieth anniversary of the Easter Rising, then demolished. This broad central street, which was called Sackville Street when O'Connell Bridge was called Carlisle Bridge, is dominated by the surviving statues, scattered like chesspieces in a finished game. The General Post Office itself, a surprisingly unimposing mass on the west side of O'Connell Street where the Easter Rising was based, still sells stamps. A small bronze statue in its central window, of a Christlike Cuchulain dying his mythological Irish death, is the principal tribute to those who were killed there. The great pillar of the Liberator of the Catholics in the 1840s, Daniel O'Connell himself, stands facing the river at the head of his street. Representative Irish figures – bishops and workers – pace around it clockwise, winged victories hover helpfully above, but O'Connell is higher than the victories. The more modest monument at the other end of O'Connell Street, where it crosses Parnell Street, is to Charles Stuart Parnell, who was not only a liberator but an adulterer, compromised at least by Victorian standards.

Between the two heroes, on the island running down the center of the broad street, is a recent addition, a fountain running over an elongated green-bronze reclining woman who is supposed to represent Anna Livia Plurabelle, James Joyce's personification of the River Liffey in *Finnegans*

Wake. Even O'Connell Bridge has a small bronze plaque on it commemorating a point in Leopold Bloom's progression around the city in *Ulysses*, one of fourteen such plaques; and Joyce himself is present as a bust in St Stephen's Green, funded by American Express, as well as a lumpy bronze figure resembling a lost tourist outside a cafeteria on O'Connell Street. All around the city stand tall glass cases displaying reproductions of prints by James Malton from his *Picturesque Views of Dublin* of 1794, so passersby can compare the modern city to its ancestor. Often the main buildings appear almost unchanged, islands of stability in a sea of people and surrounding buildings whose styles have changed radically (though I did see a horse-drawn cart proceed nonchalantly up O'Connell Street amid the cars and buses). The political figures are monuments to Irish nationalism for the Irish, but many of the other monuments seem to be less for the national audience, the audience of Parnell and O'Connell, than for the current wave of invaders, of which I was a small part.

Tourism, which brings three million foreigners to this island of three and a half million people every year, looms ever larger in the Irish economy, as it does in many other small, pretty, poor countries, from Nepal to Costa Rica. O'Connell Street itself has the buildings, monuments, and statuary of a national past, but the stores that line the street – fast-food chains, souvenir and gift shops – are there for foreign visitors. Tourists have as peculiar an effect on a culture as invaders do, if not in so straightforward a manner. They are there, officially, to see the exotic, the different, the ancient, but sooner or later a new economy springs up in their wake. Thus the culture they left behind appears again, or that limbo which is tourist culture springs up, or the place they come to see becomes its own impersonation. Sociologist of tourism Dean McCannell compares the hotel and resort complexes tourism generates in the third world to the beachheads of a quasi-military occupation – a literal invasion.

There are situations in which tourism can encourage the preservation of a place, but far more frequently, tourists inadvertently stimulate an industry at the cost of the local culture. Cultures, after all, evolve and change, but tourists most often want an unchanged vision of the past. The ultimate versions are in colonial Williamsburg in Virginia and the Irish-American Folk Village in Omagh, County Tyrone, where the past is reenacted with actors, costumes, props and sets for the audience of travelers. It's hard to say to what extent a real past has been resurrected in these places, but the present has certainly been vanquished. Such tourist accommodation raises the question of whether a tradition still exists when it's no longer carried on for traditional purposes. Thus an Aran Isles sweater knitted for an international market is not the same as an Aran Isles sweater knitted for the fishermen in the family. It looks the same, but it's part of a market economy, not a subsistence economy; subsistence and handicraft have become an aesthetic of authenticity. The vast and ever-expanding industry of tourism threatens to turn the whole world into a series of theaters whose companies perform palatable versions of their culture and history. Tourists thus possess a perverse version of Midas's touch: the authenticity and exoticism they seek is inauthenticated and homogenized by their presence.

The English call their tourist business "the heritage industry," which makes it clear that it *is* an industry and its product is the past, as it is in Ireland. The past is an odd product. In fact it's not a product at all, since it is too unknown, unpleasant, and unimaginable for vacationtime consumption; the heritage industries instead supply their audience with versions of selected aspects. The word *industry* is as peculiar as the word *heritage*, since nothing is produced but opportunities to consume and some of the necessary artifacts, from historic markers to shamrock-bedecked salt and pepper shakers, and since its purpose is leisure, which was once the opposite of industry. It is the perfect industry for the information age: one of leisure, consumption, displacement, simulation. It seems both to reverse colonialism and to repeat it; it is a means by which some of the wealth of rich nations returns to poor ones, but is also a means by which the former continue to invade and dictate to the latter.

Tourism reconstitutes as play all the endless tides of humanity that constitute war, invasion, and exile, reenacts the tragedies of population shifts as comedies of desire and expenditure. It echoes another form of movement, that of the pilgrimage, though its secular subjects are more varied and whimsical. The tourist may be in search of sun, of certain kinds of terrain and weather, of festivals, of relics and signs of the past. They are an odd species, whose anticipatory wanderings seem more satisfying than their arrivals, and it may be that the real purpose and pleasure of travel is simply not to be at home or to be in motion.

Irish tourism has its own peculiarities. The counterpoint to Ireland's history of invasions is its more recent history of emigrations, of people streaming out of the impoverished isle to populate all the English-speaking parts of the world. More than half a million of Ireland's annual visitors are US reverse immigrants, coming to look at where their ancestors came from, and they are well catered to, for tourism generates a hundred thousand Irish jobs. The tourist shops sell tiny coats of arms and other souvenirs bearing family names and crests; the National Library serves a constant flood of genealogical researchers come to put together a pedigree; most of the natives will inquire at some point whether a tourist is part Irish and will civilly listen to foggy details of the ancestral departure. The animosity between American tourists and hosts that exists in other parts of the world seems largely absent here. It may be an understanding that they are not wholly distinct peoples and their histories are intertwined; it may be respect for the cash cow; and it sometimes seemed to me that the Irish require American tourists as an audience: if they could convince the tourists to believe their version of Ireland, they might be able to believe it themselves. Tourism may theatricalize its sites, but in Ireland, many of the locals are willing actors.

One of the ineffable byproducts of these touristic transactions is sentimentality. It's hard to say how much of this sentimentality is an Irish-American view of the motherland from a pleasantly misty distance, but it's what made going to Ireland such a dubious venture for me. The place is insufferably cute in the popular American imagination, associated with shamrocks, Lucky Charms breakfast cereal, green beer on St Patrick's Day, lines of syrupy old songs like "When Irish Eyes Are Smiling," and sticky catch-phrases like "the auld sod." It's odd that a country whose history often reads as an interminable litany of the varieties of suffering should have conjured up an international image of such cloying charm, or perhaps the latter counterbalances the former – perhaps sentimentality is the pink bouquet on the coffin, as charm can be a survival skill. (And revolution and suffering have their own sentimentality. In counterpoint to the idyllic "Ireland of the Welcomes" of mainstream American fantasy and Irish Tourist Board promotion is "Ireland of the Bombs," in which simple, heroic versions of Northern Ireland and the Irish Republican Army swell up to fill the whole map of a more radical American imagination.) Sentimentality is the enjoyment of emotion for its own sake, a kind of connoisseurship of feelings without the obligation to act on them, the narcissism of the heart. Irish literature is full of sentimental drunks, and its landscape is full of tourists nostalgic for the sanitized version of the simple country life presented in postcards, on tea towels, and in vacation lodgings. Still, fine qualities go along with sentimentality, including the tenderness that is acted on.

I set off for the tomb of one of the most publicly unsentimental men in European history, Jonathan Swift, the Dean of St Patrick's Cathedral. Great currents of pale pedestrians in blues and grays and browns washed down all the major avenues, damming up at the traffic lights and swirling into each other wherever two streets met. I poured with them across O'Connell Bridge, around the bulge of Trinity College's white, many-pillared Palladian prow, up Dame Street which acquired a few other names as it wavered along, then southward, away from the Liffey, to another street of varying names that delivered me to the front of the cathedral. Empty of worshippers, if not of visitors and tombs, it seemed less like a church than a museum of a church, clean and vacant, and the woman at the door taking admission from every entrant confirmed its transformation. Both the hulking old cathedrals in Dublin belong to the Protestant Church of Ireland, as do many of the oldest and most imposing churches throughout the Republic, though their congregations constitute a tiny fraction of the population: the Republic of Ireland is 95 percent Catholic. Church and state have never been particularly separate here, and the north wall bore monuments to soldiers of the "Burmah" campaign, of South African conflicts, and of the First World War – signs that the old colony of Ireland was a launching ground for the rest of the British Empire.

Swift was a servant of the Church of England, and master of the cathedral. High above the south aisle, below a white bust of his heavy features and before the resting place of his remains, I found his famous epitaph inscribed on black marble, which Yeats translated from Swift's Latin thus:

> Swift has sailed into his rest;
> Savage indignation there
> Cannot lacerate his breast.
> Imitate him if you dare,
> World-besotted traveller; he
> Served human liberty

– an inscription which anticipates the grave's status as a tourist landmark.

Another of the reasons to travel is to situate the past in its locale, to put a picture round the facts. The particulars of a battlefield, a poet's house, a monarch's throne often illuminate dusty events, making them immediate and imaginable in a way nothing else can, or undoing the imagined version with an unexpected scale or style. It is the texture of a life that is most often missing from accounts of the past: the size of a room, the height of a border-wall, the rockiness of a landscape. For those stuffed full of the lore of a continent not our own, it's helpful to come and flesh it out once in a while with tangible places. I too was a tourist in Ireland, and I had come to look at relics of the past and literary sites, among other things. In the cathedral where Swift was dean from 1713 to his death in 1745, it's possible to picture a real man whose step covered a certain distance on the stony floor, who entered here and mounted there to preach and looked out towards that river. But most of the terrain of Swift's life has been cleared away.

His St Patrick's was situated on the lowest ground in Dublin and was surrounded by the slums that housed the city's poorest people, in narrow streets with names like Shit Street and Dunghill Court. Without sanitation or any system of waste removal at all, offal accumulated grotesquely, giving the place a stench politer eyes and noses than Swift's could hardly ignore. This filthy and often flooded little expanse was his kingdom. The dean himself once wrote sardonically of all this shit, "that these Heaps were laid there privately by British Fundaments, to make the world believe, that our Irish Vulgar do daily eat and drink." A great walker of the city, Swift was well acquainted with his poor neighbors and well loved for his charities and for his championing of their rights.

Literary historian Carole Fabricant points out that Swift's preoccupation with cruelty and injustice, with dirt and shit, has its literal ground here in the old neighborhood of St Patrick's. Swift is usually portrayed as an English writer whose misanthropy, indignation, and preoccupation with the more repulsive aspects of the body were personal idiosyncrasies, or signs of mental disorder. Though he did go mad, his themes had grounds. In Fabricant's view, Swift's harsh antiromanticism was as rooted in his residence among the poor of Ireland as his friend Alexander Pope's mannered poetry was the fruit of servants and English country-house living. Swift had an odd relationship to Ireland, once remarking, "I reckon no man truly miserable unless he be condemned to live in Ireland" and another time rejecting a return to London by saying, "I choose to be a freeman among slaves, rather than a slave among freemen." His grandparents had come over after Cromwell, and the question of whether he was English or Irish was and is answered according to desire and politics rather than any clearcut fact. It might be most accurate to say he was both. Born and raised in Ireland, he spent his young manhood in the literary and political coteries of England and the second half of his life back in his birthplace. He seems to have been something of an exile wherever he was, not wholly a member of either country, split between comfort and conscience.

Ireland has most often been defended by those who would emphasize its virtues and apologize for its failings; Swift took another approach entirely, harping on the uglinesses of a denuded, overused landscape, of poverty and powerlessness, and tracing their source to the graceful powers of England. His most famous book, *Gulliver's Travels*, universalizes his spleen to become a critique of at least European man, but the less-known majority of his work dealt with the specifics of his own time and place – pamphlets and satires on the current political situation, poems mocking the elevated motifs and ideal landscapes of his peers' conventional poetry.

English literature itself sometimes seems to me a huge country house, a mansion to which the shanties and new wings of other English-language literatures are attached; perhaps they hold up the ancient hulk at their center. In this mansion, the principal rooms are occupied by the familiar furniture of epic and lyric poetry, of the novel, the side tables and cabinets of the essay. The books I used when I was an English major folded the Irish in as English literature, but the biggest, the most central, the most familiar pieces are almost always truly English, the results of an irreplaceable confidence and centrality. There's the dark throne of Milton, the banquet tables of Shakespeare, sonnets from Sidney to Shelley scattered everywhere like bouquets of flowers, and, huge and soft and inviting, the fat featherbeds of the English novel. Swift's work sits in a passageway, a hard chair with a view through the cracks in the walls. Joyce supplied his own furnishing for the house of English literature with Stephen Dedalus's comment that the "cracked lookingglass of a servant" could stand as "the symbol of Irish art," suggesting not only a subjugation but a fractured, unpredictable reflection. Then he went and built something new, with a monument to Dublin scattered within it.

Irish writers have punctuated English literature with works that seldom rest so easily on confidence and centrality, with works that reinvent and critique the dominant forms and place the viewer in unfamiliar positions. Irish masterpieces have taken apart the conventions not only of their genres, but of narrative, language, and tradition. *Gulliver's Travels* and *Ulysses* stand at either end of occupied Irish literature, both books of mockery, exile, and wandering, by one Irishman who chose exile in Dublin and another who chose exile out of it. *Tristam Shandy*, the first and in many respects the greatest experimental English novel, was published between 1759 and 1767 by Laurence Sterne, an Irish-born clergyman, and even the Brontës, so identified with the Yorkshire moors, were raised on their Irish father's wild Irish stories. They introduced a dark, violent strain to the complacent Victorian novel – a clutch of eels in the featherbed – and other surprises were subsequently delivered by Wilde, Joyce, Synge, Shaw, Beckett. A more intricate, sardonic, restless imagination seems to characterize the Irish works claimed by England, a sensibility more cognizant of the arbitrariness of literary form and all the opportunities of subverting it.

In this century, a kind of literary geography has remapped the house of English literature. Jean Rhys's *Wide Sargasso Sea* imagined the Caribbean drama that preceded Charlotte Brontë's *Jane Eyre*; Edward Said has scrutinized the colonial rapacity that secured the stultifying calm of Jane Austen's *Mansfield Park*, whose idle gentry are living off unseen slave-plantation wealth. Early in the eighteenth century, Swift was already carrying on this cartographic operation, telling us what his gracious society looked like from behind, below, and outside. Humor itself can be a way of seeing double, of noticing the gap between how things are supposed to be and how they are, from the formal elements of logic and language to the hypocrisies of social and political life. Such an engine drives everything from a simple joke to an extended satire, from Swift's constant shift between lofty and vulgar styles in his poems to his *A Modest Proposal*, whose humor lies in its entertainment of cannibalism as a reasonable solution to Irish poverty and thereby makes apparent the well-established existence of cannibalism by other means. The most humorless are usually those who have most invested in the existing order, and humor has always been a pleasure, a tool, and a weapon of those who see that gap. The view from Dublin has often been tragic, and heroic, and sentimental, but it has sometimes been mordantly funny.

3 *Noah's Alphabet*

Later on, Dublin would become a city of people for me, but on first encounter it was a city of monuments and ideas. From Swift's tomb I steered past houses, stores, official buildings, offices toward the National Gallery, but the Natural History Museum lured me in on my way. Just after I entered its mortuary hush, a flock of children cascaded through the tall doors in a steady rush, and every fourth or fifth one uttered a high clear O of amazement, their cries falling on each other and echoing like notes of music in the huge hall. Immediately inside the doors loomed the black fossil skeletons of the three Giant Irish Deer, more popularly known as the Irish Elk. Already six feet high at the shoulder, and standing on pedestals facing the doorway, their antlers swept the air for several feet on either side like huge many-fingered hands or like condor's wings or cartoon balloons of speech, the largest antlers ever borne on earth. They looked like deer becoming birds, like those mythological hybrids of snakes and women, lions and eagles, with their avian antlers. But the children and I, after we recovered from the overwhelming, unlikely, majestic presence of the black bones of two stags and a doe, were fascinated by all the things in the museum, from the downstairs jars of worms and cases of birds to the upstairs realm of all the big mammals of the world, with its overhanging balconies of small specimens.

The kind of modern natural history museum I grew up with tries to rationalize our fascination with animals by placing each stuffed beast in a painted diorama accompanied by helpful plaques of scientific information. These displays belie our real reasons for looking at animals and push us towards a rational version of why we ought to look at them. But Dublin's Natural History Museum is itself a fossil, with displays that have hardly changed in eighty or ninety years, and the Irish Elk have been standing before the public since the 1830s. The collection doesn't try to disguise the seductions of collecting, of trophies, accumulations, abundances, and, most of all, of the forms of animals.

The whole downstairs was given over to Irish fauna, a narrowly circumscribed subject thanks to the isolation of islands – and Ireland became one fifty thousand years ago, while Britain was connected to Europe, the Thames a tributary of the Rhine, for forty thousand years more – and thanks to subsequent ice ages. It's the ice ages and isolation that did away with the reptiles on this island, though St Patrick is traditionally given credit for driving them all out. Medieval scholars – the Venerable Bede in the eighth, Giraldus Cambrensis in the twelfth century – asserted that converted Ireland was so pure and wholesome a place that nothing poisonous could thrive there and even the scrapings of Irish books could be used as an antidote to serpents' bites and other poisons. Recent scholars have speculated that St Patrick's snake-charming was an idea cooked up by the Vikings, because the saint's Irish name – Padraig – sounds like Norse *Pad-rekr*, or toad-expeller. There is only one reptile native to Ireland, *Lacerta vivipara,* a modest-looking little lizard that gives birth to live young. But the jubilantly irreverent Dr Oliver St John Gogarty, the model for Buck Mulligan in *Ulysses*, long ago loosed some snakes on Featherbed Mountain to make up for the doings of the saint, or the ice age.

The Giant Irish Deer became extinct 10,600 years ago, during another dip in temperatures, and most of the larger carnivores here have become extinct as well, too long ago to leave any taxidermic evidence. Says a sign, "The various claims for the last wolf killed here ranged from 1786 to 1810," long after bears and boars were gone, and another sign admits, "Most red deer herds have been introduced or at least managed since the thirteenth century." Which is to say that nature, in the sense in which at least people in western North America are familiar with it, as a realm perhaps influenced but not wholly controlled by human agendas, no longer exists here. People here cannot imagine themselves as visitors or guests; they are the principal tenants, having evicted more species than St Patrick was ever credited with removing.

Without many mammals to celebrate, the downstairs room, with its cakeshop-yellow walls and checkered linoleum floor, made much of its smaller species. There were lavish blue wall-cases of Irish sponges and parasitic worms grouped by their hosts – fish, cats, sheep – and bristleworms and horizontal cases of Irish moths. Most of the bottled specimens had faded to a uniform white, so that all the creatures of Ireland seemed as pale as or paler than its human population. Even an eel that had choked to death on a frog had faded with its victim, so they too seemed one peaceful pallid bottled mythological animal now, a white demon with hind legs waving like long mustachios on either side of its mouth, in contrast to the seraphic black elk. The creatures that had retained their colors were mostly demurely colored to begin with, hares and wild geese and songbirds and hundreds of indistinguishably dust-colored moths. Two nuns, the old-fashioned kind in navy-blue habits skimming the floor, approached a case of seabirds, and one of them said in a high faint voice, apropos of something else, "And it was so perfectly still."

Upstairs was all the wealth of the world, or so it seemed, a huge long hall with row after row of handsome wood-framed glass cases, replete with perfectly still dingy bone and faded fur. Here was everything from pangolins to polar bears. Trophy heads jutted off three or four sides of every pillar, and from the entrance the antlers and horns thickened the air like the branches of a winter forest. In the center, flanked by glass cases, were open platforms with giraffes, rhinos, hippos, elephants, some stuffed, some skeletons. The animals were only loosely grouped by continents and species, and there were exceptions everywhere, like the Irish wolfhound in the case of bears and wolves. Most of them appeared to be posed for formal portraits, as aristocrats once did: looking natural, their limbs arranged with carefully casual asymmetry, head held up.

Others were arranged with what seemed a peculiar symbolic purpose. A standing lion in one of the first cases had the stuffed head of another lion set on the floor between his legs, its wrathful glass eyes forever staring up at the other lion's belly and the mane around its face making it look like a dandelion snarling in a lawn somewhere. A lot of the animals were hunting trophies sent back as specimens of the hunter's prowess. There was an Indian tiger given by King George V, and a popeyed housecat, "felis domestica L shot in Donegal in 1856," whose hunter modestly remained anonymous, posed with a forepaw held up alertly, like the racehorses in George Stubbs paintings. A badly stuffed snow leopard appeared to have a face crumpled in grief, and the civet cat skeleton's head bobbed as trotting children shook the floor. "Once upon a time they were real," a father told his children, as though they had become toys or images rather than eviscerated corpses and boiled-down skeletons, a charnel house of the wilderness. There were brindled gnus, nilgais, musk oxen, and an American bison who looked like an old friend to me, part of my part of the world. One zebra or wildebeest had been caressed on the crest of its flank so often that the hair had worn off and the hide gleamed.

In this jumble was a survey of the whole natural world in terms of faded fur and yellowed bone and glassy eyes. It offered a pleasure unlike the shifty one of zoos. In a zoo one hopes to catch a glimpse of Life and often misses it, or sees prison rather than real animal life, but here one came to see Form, and it was absolutely, utterly available, overwhelmingly so. The animals served as images of themselves, like a book that had come to life, although the chambers were collections of deaths. They weren't there to be read scientifically; there was nothing about locales or habits beyond a few notes about Irish mammals. They were organized as much by aesthetics and symbolism as by taxonomic and geographic logic. They could have been read historically: like the memorials in St Patrick's Cathedral, they constituted a sidelong tribute to imperialism, to the time when servants of the British Empire covered the globe, fighting wars, dominating landscapes, and sending home curiosities. In this way they were imperial souvenirs of imperial expeditions into the larger world. Or they could be read for traces of the history of science, from the museum's own origins in an eighteenth-century scientific society in the days when science was a gentlemanly concern to the Victorian fetish for collecting and classifying the world – which, come to think of it, seems to have been a reflex of trying to put the Empire in order too.

Yet they didn't seem to be about those histories at all, but something more general, and more personal. Looking at animals lets us think about the state of being embodied, of our scale, hairlessness, bipedalism, attenuated teeth and muscle, frail bone, binocular vision, and to imagine what it would be like to be otherwise, to possess the oiled power of tigers, the stately bulk of elephants, the weightless grace of gazelles. To look at these creatures was to feel one's limbs expand and vanish, grow clawed, powerful, become fins or hooves. The great cylinders of rhino and elephant rib cages looked like baskets on stilts in which one could curl up; and there was a whale skeleton suspended in the open air above everything, where only the giraffe's head reached, as if it floated like a zeppelin or the rest of us were walking lobster-like on a sea floor. The whale had washed up in Bantry Bay on Ireland's southwestern coast in 1862. Things always look like what we have seen before, and, thanks to the sequence of the museum's displays, the whale looked like the pollywog preserved downstairs, though Jonah could have opened up a hotel in it. Later on, when I left the museum, the whole world seemed to be assembled out of forms drawn from animals: the curving double docking hooks on the quays of the Liffey looked like musk ox horns, and when I finally arrived at my original destination, the dark wooden lintels of the National Museum's many doorways looked like the antlers of Irish Elk.

Every collection is a world in miniature, but animals particularly, describing as they do the possibilities of terrain, bodily form, temperament, and danger, represent the world for us. The museum seemed like a lexicon of form, of all the forms of grace and awkwardness, the frail and the potent. The most beautiful thing there in my eyes, and one of the most beautiful I've ever seen, was a pygmy elephant skull. The two things that most distinguish a live elephant, its ears and trunk, were gone with the rest of the flesh, and the tusks stretched out like drawbridges to this massive tabletop castle of bone. Taller than it was wide or deep, it had double windows of sinuses in the otherwise blind front wall of bone, and eye sockets swinging out from the mass like balconies, so big around I could've stuck my thigh through them. Tucked deep within the bulk were a surprisingly small smile of sweet joy and a pointed chin like a commedia dell'arte mask. All the strength and intelligence of elephants seemed to be visible in the architecture of its skull, and the massiveness, the featureless frontal plane, the sideways eyes, the hidden jaw suggested how much these qualities were tied up in a corporeal condition so unlike the human. The enchanting skull of a pygmy elephant was not what I had come to Ireland to seek – but one travels for the unexpected.

Natural history museums let us imagine encountering these beasts, for at some level we still remember snakes and lions and still have reflexes that can be awakened by them. And it gives us, in their various qualities of presence, of menace and charm and power and music, characters for our dreams. Wolves survive in the imagination even in places like Ireland, where they've been extinct for two centuries, and Africa has given the rest of the world spectacular megafauna as forms for children's toys and adults' images, from the lion who lies down with the lamb to the elephant of the US Republican party. They populate our language too, though even there they are endangered.

It's no coincidence that the books and posters we use to learn the alphabet from are most often animal alphabets, from aardvark to zebra, for animals constitute the primordial alphabet. I grew up with a Dr Seuss book called *On Beyond Zebra,* which coined new letters for the alphabet and fabulous beasts to go with them, as though you couldn't have innovation in one area without the other, a proposition that made perfect sense to children. Medieval Irish manuscripts are notable for their animal ornamentation around the capital letters, as though the alphabet were turning back into beasts. Like alphabets, animals constitute a finite group that can describe the whole spectrum of possibility; animals are themselves a language for describing both the bodily forms and range of dispositions of human beings. In the Middle Ages, bestiaries were a popular form of literature, occupying a niche somewhere between field guides, fairy tales, and alphabet primers. The bestiaries, and the animals they described, were part of a system in which everything had an allegorical meaning; the whole world was a text waiting to be read by those who knew its language. Elephants, for example, signify Adam and Eve in Eden, because they are supposed to conceive their young innocently, by sharing the fruit of a certain tree; they also signify the Hebrew law, because when they fall they cannot get up again. Wild goats, because they constantly seek higher pastures, signify good preachers. The Bible and the world were two equal forms of the divine text, so that animals were almost literally an alphabet, rather as they were for Aesop, who made them illustrate so many aspects of human character and conduct, with his dogs in the manger, his virtuous ants and sybaritic grasshoppers. In either version, animals make the human world clearer, give tools and emblems with which to describe and understand it. Even as recently as George Orwell's *Animal Farm,* animals served as emblems of human tendencies, so that the horses in his allegory were honest workers, the pigs corrupt conspirators.

The majority of figures of speech that make the abstract concrete and imaginable are drawn from animals, human bodies, and spaces, from the wolf at the door to the arms of chairs and shoulders of roads to the excavation of buried memories. It's the animal world that makes being human – catty, dogged, sheepish – imaginable, and the spatial realm that makes action and achievement – career plateaus, rough spots, marshy areas – describable. Sometimes it mixes: along the Cork–Kerry coast are the jutting formations Lamb's Head, Hog's Head, Cod's Head, Crow Head, and Sheep's Head. But most of the discussion about nature and the environment emphasizes a purely physical or spiritual need for it, not its imaginative role. Not long ago, I noticed an art magazine misspelling the bridle reins of the phrase *on a tight rein* as *reign*, because although they understood royalty, they had no clue about horses and their harnesses – so even the world of domestic animals was lost to them as a way of describing the human and the phrase was becoming meaningless on its way to becoming extinct. (More recently, I found myself going to ride a horse with a few carrots and a stick as aids, and the phrase became resonantly literal again.) I wonder if generations of being without contact with such spaces and beings will eventually strip down English into a kind of newspeak. After all, how many people now know how a mule kicks, or have seen bees make beelines? And when speech goes blank, imagination will have preceded it. The Natural History Museum is a museum of language, symbol, metaphor, and imagination, of the creatures that once inhabited our lives and are now fading even from our speech.

The complete development of the world as a human-only zone – the paving over and flattening of the landscape and the elimination of all creatures but food animals sequestered in factory production sites – threatens to take away not only the imaginative solace of a world beyond us, but the very language of the mind. Metaphor is a Greek word that literally means to transport something from one place to another; and in Athens the public transit system is called the Metaphor. There one can literally take the Metaphor to work, or take the last Metaphor home, though in the rest of the world metaphors serve only as a medium of imaginative travel. They are, in fact, the transportation system of the mind, the way we make connections between disparate things, and because the connections are intuitive and aesthetic, they are the essence of the ways in which we think that machines cannot. Metaphors navigate the way things span both difference and similarity; they describe a world of both dizzying variety and intricate relationships. Without metaphor the world will seem threateningly amorphous, both boringly identical with ourselves and utterly incomprehensible. Animals, with their inherent resemblances and differences, are where metaphor begins.

The essayist John Berger writes, "The first subject matter for painting was animal. Probably the first paint was blood. Prior to that, it is not unreasonable to suppose that the first metaphor was animal. Rousseau, in his *Essay on the Origin of Languages,* maintained that language itself began with metaphor: 'As emotions were the first motives which induced man to speak, his first utterances were tropes (metaphors). Figurative language was the first language to be born, proper meanings were the last to be found.' If the first metaphor was animal, it was because the essential relationship between man and animal was metaphoric. . . . What distinguished man from animals was the human capacity for symbolic thought, the capacity which was inseparable from the development of language in which words were not mere signals, but signifiers of something other than themselves. Yet the first symbols were animals. What distinguished men from animals was born of their relationship with them." Language is humankind's principal creation, a pale shadow of Creation, and one that needs to come back again and again to the nonhuman world to renew itself, to draw strength and color. It requires contact with the natural worlds of the landscape, the body and the animal kingdom to connect its creations to Creation, and makes contact by metaphor.

The last display in this inanimate animal kingdom brought me back to Swift and his speculations on the human animal and its place on earth, or lack thereof. In the very back of the Natural History Museum in Dublin, the last case you'd come to, were four skeletons: a chimpanzee, an "Orang Utan," a gorilla, and a man. It was not a display on evolution but on comparative anatomy, a plaque hastened to clarify: "It is not to be inferred from this display that man has been derived from an anthropoid ape like those here." After all, we were still in Catholic Ireland, though the apes were from afar. Perhaps it's only that if one establishes being human as the norm, the divergences are bound to look bad by comparison, but they seemed slouching and rough here with their long pelvises and flaring bell-like ribcages beside the straight sapling of *Homo sapiens* with his pelvis like a butterfly. The apes were propped up by black rods attached to their spines and bolted to the floor, but the man was suspended from the ceiling by a golden chain attached to his skull with a wing nut. The installation seemed to propose that human and ape anatomies are analogous, but their essences are utterly different, that animals rise from the earth, but humans dangle from the heavens like God's puppets, touching the ground but disconnected from it, strangers on the earth. The installation conveyed beautifully the dissociation that distinguishes humans from animals, that causes us to create rather than inhabit Creation. On a little glass shelf above the chimp, the lacy bones of a tiny white-handed gibbon's upright and humanlike skeleton presided, like a fanged angel with arms that reached its ankles.

The suspension of the human skeleton gave visible form to what perhaps changed when the creatures who were no longer apes left the trees and began to walk upright across the land in the tenuous balance of bipedalism, their eyes focused on the distances that hardly exist in forests, their hands hanging at their sides waiting for someone or something to grasp, their voices crying across the open space. The skeleton dangled as though it belonged to the sky and needed to grow the wings most bipeds have, to lift further from the ground of its origins; or it dangled with its feet just scraping the floor of the case as though it needed to come back to earth, as though with its straight treelike body it needed to put down roots, to solidify. It seemed to me as I stood there before the glass case, grubby and jetlagged, that human beings when they became upright aspired to two conditions: becoming birds or becoming trees, wanderers or settlers, oscillating between their roots and their wings, exiled whichever way they turned.

First metaphor was animal
figurative language = first language
first symbols were animals.

4 The Butterfly Collector

Afterwards I ate a sandwich in St Stephen's Green and tried to digest the museum. The green was as mild and civilized a place as a park can be, all gracefully massed shrubbery and close-cropped lawns and tranquil waters. This place too had a violent, unimaginable history, most recently as a center for the insurgents in the Easter Rising of 1916, with the troops commanded by Countess Markievicz. She was sentenced to die for her part in the uprising, was pardoned, and became the first woman elected to the English House of Commons, though she was in jail again when she was elected. The poet James Stephens describes how on Easter Monday 1916, he came out from a quiet morning at his desk in the National Gallery to find small groups of people standing about the streets. "These people were regarding steadfastly in the direction of St. Stephen's Green Park, and they spoke occasionally to one another with that detached confidence which proved they were mutually unknown." Finally a man with a red moustache "stared at me as at a person from a different country" and explained: "'The Sinn Feiners have seized the city this morning. . . . They seized the city at eleven o'clock this morning. The green there is full of them. They have captured the Castle. They have taken the Post Office.'

"'My God!' said I, staring at him, and instantly I turned and went running towards the Green.

"In a few seconds I banished astonishment and began to walk. As I drew near the Green rifle fire began like sharply-cracking whips. It was from the further side. I saw that the gates were closed and men were standing inside with guns on their shoulders... In the centre of this side of the Park a rough barricade of carts and motor cars had been stretched. It was still full of gaps. Behind it was a halted tram, and along the vistas of the Green one saw other trams derelict, untenanted." Easter Monday had been set aside for leisure by everyone but the small army of rebels: Stephens was teaching himself to read music.

While I ate my sandwich of egg and marvelous bread in St Stephen's, flocks of chickens and the usual ducks of city parks hunted for crumbs along the banks of the pond. People sat in the weak sun or strolled, themselves so mild and civil it was as hard to imagine them kin to the tough fighters of the time as to picture the lush trees and lawns of this park interspersed with barricades and desperados, punctuated by gunfire.

*

I had found something else in the Natural History Museum I had been looking for outside. I'd come to Ireland fascinated and impressed by Roger Casement, who had been instrumental in the Easter Rising and who was hanged for treason a few months later. He was among the most thoughtful of Ireland's heroes, and so complex a character that I was foolish to expect some bronze or marble tribute to him in the streets. Instead I found what seems to be his only monument, in a glass case on the ground floor of the museum, protected from light by a soft imitation-leather cover, so the case had to be opened like a book. In this case, at the beginning of a row of similar covered insect cases, was a huge tropical butterfly all alone, surrounded by poetry on the subject of butterflies. With its deep orange wings bordered in black, a white spot at their upper ends, and a pin through its heart, it hardly looked the worse for age. "A South American butterfly collected for the Natural History Museum by Sir Roger Casement circa 1911," read the inscription on this frail monument.

The reasons why Casement was in South America in 1910 and 1911 have everything to do with his eventual involvement in the uprising, though only circuitously, for Casement's was a circuitous life, one that opened up meanings and histories with which England and Ireland still haven't come to terms (and most of his biographers have openly disliked him in a way almost unique in the genre). Casement dangled between two worlds for most of his life, two countries, two churches, two philosophies, between the respectable and the revolutionary in both his private and political lives, exiled no matter which he chose. It seems built into his name, for a casement is a window, something that itself mediates between two realms and is contained by neither. His biography is the tale of the evolution of a good imperialist into a great anti-imperialist, of a man rewarded and then terribly punished for following his principles to their logical conclusion, a tale scrawled across four continents and half a century. Yet for all the reach of his life, from Ireland to Africa and the Amazon, around Europe and the US and back, his real territory of exploration was closer to home, or closer than home. The Empire of the Body, of Pain and then of Pleasure, was the true subject of his reports back to his culture, the body as the greatest unknown, the region of taboo and mystery, the darkest continent only beginning to be explored and mapped.

The most illuminating of the few stories about his childhood describes how his father would swim out to sea with one of his three sons on his back and then leave the child to his own resources. Roger Casement, the youngest son, was the first to learn to swim and remained a strong and enthusiastic swimmer all his life, swimming even in the crocodile-infested tributaries of the Congo. And perhaps his father's aquatics prepared him for all the other abrupt drops into unfamiliar and foreign places that would make up his adult life, a life in which he was always somehow out to sea, finding his bearings in dangerous surroundings, perhaps even choosing to live where he would be no more out of place than the other Europeans and uprooted locals.

Nominally an Irish Protestant from the northern county of Antrim (which includes Belfast), born into a tradition of Casements who served the British Empire as soldiers, Casement had an Irish Catholic mother and was secretly baptized a Catholic as a small child. She died in 1873, when he was nine, and his father followed suit four years later, leaving the four children penniless dependents of their cousins. Not much else is known about his childhood, though he did leave us a glimpse of his education in a letter responding to his old school's request for donations. "I was taught nothing about Ireland at Ballymena School. I don't think the word was ever mentioned in a single class of the school and all I know of my country I learnt outside the school. I do not think that is a good or healthy state of mind in which to bring up the youth of any country – and while it endures, as it unhappily does, in so many of the schools in Ireland – which are in but not of Ireland – we shall see our country possessing inhabitants fit to succeed and prosper in every country but their own – citizens of the world, maybe, but not of Ireland."

Casement himself went on to become a citizen of the world first. At seventeen, he went to work as a clerk in a Liverpool shipping company dealing mostly in West African goods, and when he was twenty, he got to Africa. The great waves of invasion, conquest, and colonization that had swept over the Americas and Asia had turned to Africa later, and when Casement came in 1884, the most powerful nations of Europe were dividing up the continent like greedy children tearing apart a cake. The explorations which were to become so fixed an element of boyish adventure stories began largely as surveys into what could be claimed; Casement joined one and seems to have traveled and grown familiar with the terrain and indigenous populations of Africa. But little is known of these early years either. Joseph Conrad, in the course of his adventures which would become the novella *Heart of Darkness*, met him in the Congo, and they spent three weeks together. As Conrad recalled later, with a novelist's disregard for strict fact, "I have seen him start off into an unspeakable wilderness swinging a crookhandled stick for all weapon. . . . A few months afterward it so happened that I saw him come out again, a little leaner, a little browner, with his stick, dogs, and Loanda boy, and quietly serene as though he had been for a stroll in the park. . . . I always thought some part of Las Casas' soul had found refuge in his indomitable body. . . . He could tell you things! Things I have tried to forget, things I never did know." And Conrad was right; he never did know them, for the exiled Pole who became a patriotic Englishman never understood imperialism as Casement did, or as did Father Bartolomé de las Casas, the great champion of indigenous people's humanity and rights in sixteenth-century Latin America.

After knocking about as an elephant hunter, traveling to the United States, and revisiting Ireland, Casement returned to Africa in 1892 as a representative of the British government in what was then called the Oil Rivers Protectorate and is now known as Nigeria. For two decades, with interruptions for reasons of health and conscience, Casement was in the foreign service. A tall, slender, dark-haired man with a melodious voice, neat beard, and, in nearly all his photographs, air of noble melancholy, he was immensely handsome in the way that is usually called distinguished. Casement had an enormous tenderness for the powerless and suffering and a more private tendency to resent and suspect the authorities with which he worked – sometimes with good reason. Alternately brilliant and obtuse, a writer of poetry and prose that could be turgid, florid, and sentimental, an occasional concocter of harebrained schemes, he yet perceived things few others could at the time. "Thinks, speaks well, most intelligent and very sympathetic," Conrad had noted in his diary upon meeting Casement. Courage and kindness were the most unwavering elements of his character.

Sometime around the turn of the century, Africa began to change his ideas. In the Boer War, Casement played a role as a watchdog in a nearby region, and though he discovered no arms shipments taking the roundabout route through Portuguese East Africa, he found doubts about the government he served. "I had been away from Ireland for years, out of touch with everything native to my heart and mind, trying hard to do my duty and every fresh act of duty made me appreciably nearer to the ideal of the Englishman," he wrote in one of those retrospective letters that shed what little light there is on his life before 1903. "I had accepted Imperialism. British rule was to be extended at all costs, because it was the best for everyone under the sun, and those who opposed that extension ought rightly to be 'smashed.' I was on the high road to being a regular Imperialist Jingo. . . . Well the war gave me qualms at the end, and finally when up in those lonely Congo forests where I found Leopold, I found also myself, the incorrigible Irishman."

What hadn't been apparent to Casement in Ireland itself became evident by proxy in the other colonies of the British Empire, and when he became an Irish nationalist, he was unique in his evolution. Most of the Irish revolutionaries became so through a passionate devotion to their own homeland as distinct from all others, a provincial love rather than a general matter of principle. There are two bases for a struggle for liberation. One proposes that there are universal human rights that bear on the situation but extend beyond it; the other, that the oppressors have made a simple error in targeting a particular group with the otherwise legitimate tools of repression or extermination, rather than recognizing that group's merit. The latter belief can lead to movements of self-liberation that don't proceed by analogy to the larger population, the general principle. The great Irish revolutionary Wolfe Tone, for example, both before and after his failed uprising in 1796, fondly promoted a scheme for England to establish a military colony in the South Seas to "put a bridle on Spain in times of peace, and to annoy her grievously in times of war." Casement was perhaps the first to see Ireland as a colony much like the farther-flung and more recent colonies, and he came to understand his own country's situation by analogy with that of the Congo and the Putumayo of Peru. To identify Ireland with other European conquests rather than other European nations was, at the time, a great leap of unprejudiced insight. Casement later wrote, "It was only because I was an Irishman that I could understand fully, I think, the whole scheme of wrong-doing at work on the Congo," but the sequence of events suggests it was only because he witnessed the excesses of empire-building in the latter that he could see Ireland the way he did.

The word *imperialism* is tired now and no longer reverberates with the power of emperors and conquests and fantasies of a superior way of life spreading across the globe, but in Casement's day it still had the capacity to stir. Queen Victoria had been declared Empress of India in 1876, and by the early twentieth century a few European superpowers controlled much of the rest of the world and asserted that this was a good and an inevitable thing. Imperialism most obviously meant the conquest and control of other nations for economic and strategic reasons. It also meant that the conquerors themselves regarded and instructed their imperial subjects to regard the colonized countries as the outskirts rather than the center of the world, literally marginalized them as poor imitations, provincials, barbarians, back yards, outbacks. The consequences of this relocation and devaluation of regional identities are still being felt, perpetuated by the traditional beneficiaries, dissected by outsider scholars, reversed by the postcolonial nationalisms and nativisms which have sometimes liberated former colonies and sometimes created new regimes of authoritarianism in them. A nationalist movement or an independence movement is an assertion by its promulgators that theirs is a center, not a periphery.

In 1900 Casement was stationed in the Congo again, where he performed the investigation that made his first, heroic reputation. Leopold II of Belgium had played the French and English against each other so cleverly that he was able to annex a huge region around the Congo river, not for

Belgium, but for himself, so that he was not a king reigning over citizens with all the traditions of mystical duty and connection that still meant, but a propertyowner administering tenants. He never visited the Congo Free State that Europe recognized as his own; it seems to have been nothing more to him than an intangible from which the abstraction of wealth could be wrung. By the time Casement returned to the Congo, the place was famous in some circles for the brutal means by which the native population was made to turn Leopold's profit, and in 1903 Casement was able to convince his superiors to let him investigate the situation.

Casement's report on the Congo was published in 1904 as a Blue Book, an official government document, after the names and places he gave were removed, lessening its impact. It is still a devastating tract. The whole region of some 900,000 square miles was being administered as a work camp with such brutality that it was rapidly being depopulated. The natives who weren't killed outright were being starved, beaten, tortured, and worked to death. A bureaucracy had been set up by Leopold's officials which was permitted to levy a tax on the natives, usually of raw rubber. One such worker told Casement, "It used to take ten days to get the twenty baskets of rubber – we were always in the forest to find the rubber vines, [we had] to go without food, and our women had to give up cultivating the fields and gardens. Then we starved. Wild beasts – the leopards killed some of us while we were working away in the forest and others got lost or died from exposure and starvation and we begged the white men to leave us alone, saying we could get no more rubber, but the white men and their soldiers said: 'Go. You are only beasts yourselves, you are only nyama [meat].' We tried, always going further into the forest, and when we failed and our rubber was short, the soldiers came to our towns and killed us. Many were shot, some had their ears cut off; others were tied up with ropes round their necks and bodies and taken away."

Casement's report is numbingly repetitive, always in close-up on the suffering and death inflicted, with mutilation and murder on almost every page. There are stories of how hands were cut off as a means of proving that a punishment or death had taken place, and soldiers came in with baskets full of hands. The soldiers who administered these crimes were often Africans themselves, but the orders came from the Europeans at the top. It is still sometimes said that Leopold's administration did a good thing in doing away with the Arab slave trade in the region, but the conditions under which the people there lived seem far worse than slavery, for a slave is at least a valuable possession, while the inhabitants of the Congo were being exterminated profligately. As an economic strategy, abusing the population to mass death was bizarre, and yet it was not ideological in any way that has been identified.

Elaine Scarry, in her study of the functions of pain, writes, "Brutal, savage, and barbaric torture self-consciously and explicitly announces its own nature as an undoing of civilization, acts out the uncreating of the created contents of consciousness." Which is to say, the Belgians were, in her terms, undoing the Africans' consciousness and culture while announcing that these Africans had never possessed such things. Torture, pain, injury, captivity, hunger, and fear in their immediate effects so fill the mind that they crowd out the ability to think of other things, to remember; in their longer consequences they destroy the ability of people to perpetuate their culture by the generation of sustenance, transmission of stories, practice of traditions, and raising of children. It could also be said that Leopold's administrators were losing their own culture in their attempts to destroy that of others or revealing the undersides and limits of that culture. They seemed to be trying to brutalize Africa into their image of it, as chaotic, full of bestial, idiotic, lost people in need of an order that could only come from outside. They were ultimately producing another first-world fantasy: the uninhabited "virgin" wilderness that still exists in many wildlife and nature documentaries about the former colonies from Africa to the Arctic and in the fantasies of would-be discoverers then and now. Casement himself noticed how drastically the population had diminished since his early travels along the Congo; he estimated the regime killed three million people; later estimates reach as high as six million, the size of the Jewish holocaust under the Nazis.

Leopold had written in 1897, "The task which the State agents have to accomplish in the Congo is noble and elevated. It is incumbent upon them to carry on the work of the civilization of Equatorial Africa. . . . Civilized society attaches to human life a value unknown among savage peoples." It was Casement's novel task to restore voices to those who were being rendered voiceless, to give an authority to their experience that would undo the legitimacy of the authority of imperialism itself, and to do so by making the private and individual and local pain of their bodies a public, political issue thousands of miles away in Europe and America. Casement's report seems to have been one of those watersheds, like Auschwitz or Vietnam, which force self-doubt upon those who conceive of themselves or their culture as civilized. Although his information was not wholly new, it was given for the first time with an exhaustive detail and a credibility that could not be ignored. It seems to have been Casement more than any other figure before the First World War who delivered the first blows to the fulsome self-confidence that had spurred the Victorians to swarm over the globe. The doubts and schisms that so marked his own life would become hallmarks of the coming century, and the Congo report is one of its earliest milestones.

Casement's reports, which made him an internationally famous hero, also complicated the position of his superiors, who were not entirely enthusiastic about his work. He himself realized that they were not profoundly different from Leopold and his minions in the Congo and that their interest in his report was in its potential to compromise the Belgian ruler and increase their own power, rather than to champion human rights. His own position as a consul seems to have been a situation of convenience rather than conviction for him, a good place from which to operate, for a while. Leopold challenged his report, and some saw it as Protestant Britain's attack on Catholics – an Irish-American newspaper was among the accusers. But the independent commission Leopold appointed to exonerate him instead corroborated Casement. In 1908 Leopold was forced to hand over his private empire to the state of Belgium. The Belgian Congo achieved independence in 1960 and is now Zaïre; since 1965 it has been ruled by President Mobutu, a US-supported dictator who has exploited the tropical land and people so effectively that he has become one of the richest men in the world. The Congo river remains the principal means of transportation in this country a quarter the size of the US and still short on roads.

A few years ago, my friend Hilary traveled down the route that Conrad and Casement had traveled up, and the Congo she encountered still seemed infested with the practices unloosed by the Belgians. It was an incredibly lush, violent, backward country, she told me, and the river transport was simply a string of barges lashed together with about three thousand people on board. She had reserved a cabin with some other travelers she met, but when they went in, the floor was covered in maggots, so they used it to store their baggage and slept in the corridors like everyone else. There was an abundance of life, life everywhere, crocodiles in the river as Conrad described them, and the jungle overtaking the colonial city that had been Stanleyville and is now Kisangani. At the second village, a huge shipment of bales of smoked monkeys was taken aboard, the fur still on them and their faces frozen in grimaces. I asked if smoked monkey was a delicacy or a staple there, and she said that *food* is a delicacy in Zaïre. The army was everywhere. One day she and the other foreigners were sunbathing on deck when some of the soldiers on the ship struck up a conversation with them: they remarked that the travelers were lying in what was a favorite torture position of theirs. Where did you learn how to torture, asked Hilary, and they answered, In the United States. Whatever Casement accomplished, he built no eternal bulwarks against inhumanity, and in the long run nationalism has not always proven to be kinder than imperialism.

After his Congo report, Casement became a new kind of hero, marching as of old into the jungles of empire, not to expand Europe's hold on them but to contract and reform it. He thought of quitting the foreign service, took time off to recover his health, and traveled in Ireland. It was then his nationalist consciousness began to awaken, partly because of his growing friendship with the writer Alice Stopford Green, herself an ardent nationalist and historian. Casement decided that despite his mixed ancestry, his lot and his loyalty lay with Ireland, not England. The Gaelic Revival, an endeavor which began as an effort to save the native language of the island from extinction and ended as the revolution Casement would join, was in full swing. The tenor of the time appears in James Joyce's "The Dead," in which a young woman rebukes the story's protagonist for taking his vacations abroad when he could be learning Irish in the west; nationalism both political and cultural was coming into its own. Among Casement's papers from this time is an acerbic note from his bank asking him to please not correspond in Irish.

For Casement, Ireland seems to have functioned more as an ideal home, a ground for identity, than as a place that could contain him; for all his nationalism, he continued to spend most of his time far away, and some of his remarks suggest that it was in Africa that he felt most at ease, perhaps because colonial Africa was itself as between definitions as he was. Unable to find other work, he returned to active duty in the foreign service, was posted to South America, and found his own way to the Putumayo, a region of what was then Peru and is now the Peru/Columbia border, named after the Amazon tributary river of the same name.

The Putumayo, where Casement caught the butterfly I came across in the Natural History Museum, was essentially a rerun of the Congo, though the results of his Putumayo report weren't as dramatic. Like the Congo, it was a rubber-tapping region turned into a private slave-labor camp. His 1910 journal of the expedition is an odd mix of subjects jotted down casually. "September 30th . . . the new method of torture being to hold them under water while they wash the rubber, to terrify them! Also floggings and putting in guns and flogging with machetes across the back. . . . then sent for Francisco and will interrogate later tonight. I bathed in the river, delightful, and Andokes came down and caught butterflies for Barnes and I. Then a Capitan embraced us laying his head against our breasts, I never saw so touching a thing, poor soul, he felt we were their friends. Gielgud must be told to stop calling me Casement, it is infernal cheek. Not well. No dinner." On 6 October he noted splendid Emperor butterflies, and on the next day, "magnificent display of butterflies; beats anything I've seen yet." On 27 October he caught three butterflies on the road, and an expanded diary notes, ". . . to relieve our feelings we began an elaborate butterfly chase there & then on the sandy bank of the river. They were certainly magnificent specimens & the soil was aflame with glowing wings – black & yellow of extraordinary size – the glorious blue & white, and swarms of reddish orange, yellow-ochre, gamboge & sulphur."

One of his biographers says that the butterfly expeditions were a way to hear evidence out of reach of the overseers. The butterflies, the annoying traveling companions, unavoidable dinners with murderers, his own ailments, his many swims, his admiring looks at nearly nude natives: none of this is part of the official report. Like the other, it is a relentlessly detailed account of the varieties, locales, and inflicters of torture, the political information sifted out of all the range of his interest in the jungle. Like the Congo report, this one portrayed a brutality that was supposed to enforce an economic program of rubber harvesting, but was in fact eliminating its workforce – "I said to this man that under the actual regime I feared the entire Indian population would be gone in ten years, and he answered, 'I give it six. . . .'" Casement considered its horror surpassed anything he had seen in Africa.

Picture the enormous weight of Casement's responsibility to his government, his conscience, the Putumayan people his heart went out to all around him, the weight of suffering and death; picture the tropical leaves, the mud and the humid air, a world in which gravity must have pressed down like that of some vast, strange planet, and amidst it all the weightless airy rambles of the butterflies. Theodor Adorno once said that after Auschwitz there could be no poetry; should there be butterflies amidst atrocities? A perennial question for revolutionaries and activists is whether they should themselves enjoy the pleasant fruits they are trying to secure for others. Casement's answer is affirmative; there should be sapphire and sulphur-colored butterflies to chase and rivers to swim in and journals to keep, for the interminable task of fighting for justice demands its moments of reprieve. When Adorno spoke, his generation imagined the holocaust inflicted upon Jews – and Gypsies, homosexuals, and dissidents, among others – as unique, having already forgotten Cromwell in Ireland, the Turks in Armenia, and Casement's reports and not foreseeing the Cambodias, Guatemalas and Rwandas that lay ahead. There were poets *in* Auschwitz, writers like Primo Levi, who could quote Dante inside the camps and who survived to write his own lyric, damning books. Casement's butterflies seem to propose the complexity, the irreducibility of experience even at such terrible moments. When T. F. Meagher, a leader of Young Ireland's 1848 revolt, thought of its momentary triumph afterward, in his exile, he found it impossible not to recall as well the hair of the women in the hilltop crowds of supporters, "disordered, drenched, and tangled, streaming in the roaring wind of voices." Maybe butterflies and atrocities, like victories and streaming hair, are inseparable in memory and experience, however sifted out by reason.

In 1910, Casement was in his midforties, accelerating in action, in sense of purpose. In 1911, the government that would kill him five years later knighted him for his efforts on behalf of human rights, an honor he seems to have regarded dubiously. He was already separating himself from England, as was much of Ireland. In 1912, Irish Home Rule was introduced into the British Parliament for the third time since Parnell had unsuccessfully first raised the possibility of dissolving the political union made in 1800. The reaction in the north, fanned by the demagogue Edward Carson, was extreme. The Protestants who had come there generations before as colonizing economic and religious refugees and who in Wolfe Tone's time had been the strongest advocates of independence, began to organize as unionists. They began collecting weapons and forming a volunteer militia, which soon numbered more than a hundred thousand men. They were in the peculiar position of threatening armed struggle against the government should it cease to govern them. Afterwards, in the south, the National Volunteers were founded as a counterforce, though they had a much harder time getting arms into the country. As tension increased in Ireland, incidents of police brutality began to rouse public sentiment against the government.

By the time the First World War broke out in August of 1914, Casement was already in the United States, trying to tap the vast immigrant Irish population for funds for guns. Casement essentially failed at his labors on behalf of the revolt; the Americans did not give nearly what he hoped for, and he spent the year and a half before the Easter Rising as a self-appointed envoy in Germany miserably trying to recruit Irish prisoners of war for a liberation army and to wring support out of the indifferent German government. Often sick, sometimes paranoid, and conscious of his futility by the end, Casement had made a huge mistake when he presumed that his enemy's enemy would be a friend. Sadly disillusioned, he managed at least to get a shipment of weapons (sunk off the Irish coast) and submarine transport for himself to return to Ireland on Good Friday 1916, the eve of the Easter Rising. He hoped to tell them the support they hoped for would not arrive and therefore to postpone the uprising.

For the teacher and Gaelic revivalist Patrick Pearse, who was the kingpin of the rising, the meaning and power of the event was to be more symbolic than literal, an Easter bloodshed that would launch Ireland's resurrection – by capturing not Dublin itself but the imaginations of the Irish. The imagery is steeped in Christianity – what else can you call an Easter Rising? – and blood sacrifice is part of the redemption. The proclamation Pearse read from the steps of the General Post Office is as much a part of the poets' community it grew out of as the Declaration of Independence is Enlightenment literature. Ireland "summons her children to her flag and strikes for her freedom," as though the rebels were only carrying out a mystical mandate generated elsewhere, by Ireland itself, an island possessed of gender and intent. The embodiment of Ireland as a woman, young, ancient, or maternal, has a long and often sentimental history, but its appearance in the opening sentence of a revolutionary document is noteworthy. It calls attention to nationalism as a cause rising out of a common grounding in tales and ballads, and it suggests how much this revolution had originated as mythology and poetry and sentiment itself: the Gaelic Revival. It was a revolution made on poets' desks, in summer language schools, in painstaking research into old bards and older myths, a revolution of the imagination which laid the groundwork for independence when it reinvented Ireland as a proud and distinct country, not a poor dependent British isle.

A revolution wrought by people slowly mastering the intricacies of Irish grammar is as marvelous as the butterflies of Casement's Putumayo, unexpected, over the top. Perhaps it's the weight and weightlessness of poetry itself at work. A national uprising depends on a national identity; and nowhere is the political weight of poetry more evident than in the flowering of Irish culture that culminated in the Easter Rising, whose crucial declaration was signed by three poets, two teachers, a musician, and a labor organizer who was also a historian. Casement evaluated only the uprising's literal, immediate results and underestimated its intangible, symbolic possibilities. It was his greatest error, a politician's rather than a poet's calculation. He only succeeded in canceling the islandwide uprising called for Easter Sunday, turning it instead into a Dublin-only Easter Monday beginning. Who knows now what would have happened had he understood, or failed to communicate at all, or never landed?

Casement seems to have been at sea his whole life, where his father left him, among foreigners and enemies, equivocal in his loyalties, his private life at odds with public practice. When the German submarine sent him and two others ashore near the Dingle Peninsula in Ireland's west, he at last came to earth, and to himself, as a nationalist and a martyr. His description of his arrival in a little coracle, from a letter to his sister, is among the best things he wrote: "When I landed in Ireland that morning . . . swamped and swimming ashore on an unknown strand I was happy for the first time for over a year. Although I knew that this fate waited on me . . . I cannot tell you what I felt. The sandhills were full of skylarks, rising in the dawn, the first I had heard for years – the first sound I heard through the surf was their song as I waded in through the breakers, and they kept rising all the time up to the old rath at Currsahone . . . and all round were primroses and wild violets, and the singing of the skylarks in the air, and I was back in Ireland again." He knew that he was coming to join the Easter conspirators in death, and accepted it calmly when he was caught; in fact, he seems to have been calmer from his landing to his death than through most of the years before.

Three drenched strangers could hardly pass unnoticed during wartime in the remote region they'd landed in, and they were soon hauled in by the local constabulary. Casement at first claimed to be an Englishman from Buckinghamshire who'd written a life of St Brendan the Navigator. He later managed to get word to the rebels through a local priest before he was transferred from Tralee across Ireland – he passed through Dublin without knowing what was happening there – to London. Told he would be charged with high treason, he said, "I hope so." By the time of his trial, the seven signatories of the Proclamation had been executed by firing squads. Wartime England treated Casement with medieval ferocity: he was imprisoned in the Tower of London at first, under military supervision and the constant glare of lights, in the clothes that had dried upon him in Kerry.

Ironically enough for a charge of treason, the law by which he was tried dated back to 1351 and was written in archaic French, a relic of the time when England's rulers still had roots and claims in France and little reach in Ireland. Much of the deliberation about whether or not he had committed treason depended on whether the text implied a comma and how a word should be translated; some of the local witnesses to his arrival in Kerry spoke in a brogue the English court could hardly understand. The ideas of nationhood behind the charges were undermined by such language problems; and Casement's fellow Irishman George Bernard Shaw thought he should have put up the defense that he was not an Englishman and could therefore not commit treason against England. He did so, but only in his final speech after the death sentence had been read, a speech of surpassing clarity and integrity.

The most damning evidence produced against Casement had nothing to do with treason and never appeared in court, and yet more than anything it sealed his fate and shattered his reputation. The police had seized the papers Casement had left in London, including diaries written in 1903, 1910, and 1911, diaries which laconically document his very active sex life with other men. Casement's trial took place only twenty-one years after Oscar Wilde's trial for homosexuality, which ended in conviction, prison, then exile, disgrace, and, in Wilde's words, "dying beyond my means." Although Casement was not in court for the same reason, his sexuality was held up to be judged by a public that had changed little since his fellow Irishman's fall from public grace. With his many influential friends and heroic past, Casement might well have been pardoned, but transcripts from the diaries were widely distributed to the press and influential members of the public as a smear campaign to undermine any such support. Some of his friends deserted him, some stood fast, and the "black diaries," as they became known, were widely regarded afterward as forgeries. Many in Ireland still recalled the smear campaign against Parnell: forged letters suggested he had condoned the infamous Phoenix Park murders of the secretary and undersecretary for Irish affairs. He was cleared (though he was later accused of adultery, a genuine charge that ended his career). Books have been written to prove that Casement too was the victim of trumped-up evidence, but it seems clear now that the diaries are genuine. What has changed is the ways they can be read.

Poor Casement, condemned to be a window onto a closet. It seems to be his sexuality that renders most of his biographers so hostile. They continually snipe at him for all kinds of trivial things, from lack of landscape appreciation to sex with members of "the lowest orders." One even claims, in the wake of the Anthony Blunt spy scandals at the end of the 1970s, that there is apparently a high correlation between treason and homosexuality, which would be more credible had Blunt and his Soviet spy circle been the only gay men in England during the Cold War. The most authoritative of the biographers claims in his closing paragraphs that Casement's "nature was divided to a depth just short of real pathology, of disastrous incoherence." Yet little more than overwrought emotion and bad writing can be laid at Casement's door (and at the doors of most late Victorians); his courage, kindness, and commitment to the rights of the oppressed make the course of his life coherent and transform the apparent about-face from civil servant to revolutionary into a logical progression from redressing imperial cruelty in Africa and the Americas to attacking it in Europe.

Pleasure and pain are among the most incommunicable of experiences, the most private, the most shameful, and taboo. Casement described them in terms of their objective effects and outward signs. In his official reports he gives us the descriptions of mutilations, decimated populations, suppurating wounds from beatings, severed body parts, and corpses. In his private journals, his erotic encounters are reduced to a few descriptive phrases of beautiful eyes, large cocks, and sexual acts. The reports were official political documents, but why did Casement keep the diaries, which were to be devastating, perhaps even fatal, for him? It may be that they represented an outlet in which he could acknowledge a life that had otherwise to be kept a deep secret, or that they represented a means of keeping track of his full and busy life: weather, social engagements, bodily health, and financial transactions are equally documented in these ledger diaries. They are not the lengthy literary diaries of so many Victorians, but short notes to help him sort out and perhaps reconstruct his busy days.

Or perhaps he too was a collector in that Victorian tradition, a collector of his own experiences. The way he describes the men he desires and his encounters with them almost resembles an entomologist's notes – and entomologists have passions and pleasures too. In fact the constant harping on measurements – in his case of penises, in heterosexual men's accounts often on other quantifiable zones and scores – reminded me of fishermen and athletes, whose triumphs also tend to emphasize the quantifiable. Perhaps it is appropriate that Casement should be monumentalized by a butterfly, in a museum that is itself a monument to the variety of the world, to the imperial project of the British Empire, and to the desire to collect and classify. It is a last ironic footnote that the word *butterfly* in Spanish – *mariposa* – is a Latin American slang term roughly equal to *faggot* or *fairy.*

To read Casement's reports and journals is to recover a sense of how limited the possibilities of the body were supposed to be. The Victorian body, conceived of as incapable of inflicting and experiencing torture or other than narrowly defined heterosexual pleasures, seems a curiously wooden, inert realm, the dark continent that they were avoiding with their excess of action on every other front. Casement's sexuality sheds a different light on his rootlessness, suggesting that perhaps those other cultures far from Ireland were better locales for him to come to terms with and live out his desires. For a San Franciscan such as myself, the 1903 and 1910 diaries, which were published by Grove Press in 1959, seem quite mild, hardly as experimental or explicit as what one might read in the gay men's classifieds in many big cities. What stands out is that at a time when almost any kind of sex was laden with shame and a sense of dirtiness, Casement appreciates his partners and enjoys himself with them, be they longtime lovers or casual pickups of any race. Compared to, say, Henry Miller, who two decades later wrote so publicly of his own casual and commercial sexual encounters, he is admirable for neither boasting about himself nor denigrating his partners. It seems peculiar from this perspective that a culture which could bear Casement's revelations about torture couldn't stand his private narratives of pleasure; and it may be his sexual identity as much as his Irishness that let him see things from an unprecedented perspective, let him recognize the effects of power from a point beyond the range of its magnetism, and let him give voice to what had been silent.

Homosexuality undoes the simplicity of gender roles, leaving us with only a few possibilities: either that the conventional roles themselves are in fact social conventions narrowing down the broad spectrum of possibilities for each gender or that any divergence from the sanctioned roles is a crime, a disease, an aberration. Either gay men and lesbians are not normal or such normalcy is a majority fiction. Male homosexuality is particularly threatening to the status quo. That men can be objects of desire as well as possessors of it, that they can be penetrated as well as penetrating, undoes the unilateral imagination of the dynamics of power and gender. The response to Casement's sexuality suggests that masculinity was even more primary an element of identity than race and empire, and if it could be unsettled, everything might be open to redefinition. Casement, with his reports and his liaisons, was undoing authority at both ends of the spectrum, in the Empire and in the bedroom. In his labor and pleasure he reinvented men as various creatures, capable of many configurations, of more cruelty, more vulnerability, more possibility. Ireland has since its revolution become famous for sexual conservatism promulgated by the church and state, and most gay men there nowadays, a lesbian poet from Dublin told me, are still as closeted as Casement was.

First Casement described the world of torture and became a hero, and then he described the world of erotic pleasure and became a villain, but both his reports shook up his auditors' worldview. In the Grove Press volume on Casement, the reports and diaries are printed side by side, two versions of reality. The official, left-hand one leaves out time and locale to provide a dry summary of the testimony encountered and the conclusions reached; Casement appears only as an eye and an ear judging the material at hand. But the right-hand pages give us the whole body of unsorted experience, of eyeing handsome native men, interrogating officials and survivors, mixing with expatriate Europeans, tending his own mild ailments, recording the rise and fall of rivers, swimming, chasing butterflies, having sex, and admiring the scenery. It is odd, and honest, this mix of activity.

Casement converted to Catholicism on his deathbed and went to his death, he said, with the body of his god as his last meal, seeming at last to understand the symbolism of sacrifice around which Pearse had organized the Easter Rising. He was hanged on 4 August 1916, victim of the same violence against the bodies of empire he had spent so much of his life exposing. His own corpse was tossed into quicklime, so it disintegrated while his tropical butterfly lived on, dry and remote, in the Natural History Museum, evidence of an interlude in the rainforest when he left off chasing torturers to pursue the visions of dazzling color he describes. His remains were exhumed in 1965 for a state funeral and burial in the private plot his sister had bought him forty years before, his tiny allotment of the land he helped liberate, or didn't. He inspired Yeats to write two poems, one assuming that the diaries were perfidious forgeries, the other seemingly partaking of Casement's melodramatic style itself, with its refrain, "The ghost of Roger Casement is knocking at the door." He had jokingly promised to return as a ghost in clanking armor contemporaneous with the statute that condemned him, and in the 1970s it was reported that Casement's ghost did make frequent appearances – but in Calabar, Nigeria, where he had been a consul at the beginning of his involvement with British imperialism. "The apparition was always said to be of a kindly nature," the Nigerian report added.

5 *The Beggar's Rounds*

It was scenery that would hardly interrupt a dream, and I kept waking up to glimpses of green fields with square ruined towers in the distance and little towns where everyone seemed to have a purpose but me, then drowsing off again on this busride from Dublin to Cork. Halfway between the two cities, the bus stopped for an hour at a pub, and all the passengers dismounted and drank tea or beer, the twin elixirs which pour forth in such abundance to fuel and modify the national temperament. By early evening we were in Cork. It didn't feel like a city at all: all its industry (and that industry's infamous pollution) was out of sight from the center of town, and that center was nothing but a long curving main street that scraped up against the big university and then petered out.

More and more students are choosing to stay in university through graduate school, one of the University of Cork graduate students told me in a pub called The Thirsty Scholar, as a way to keep out of the abysmal job market as long as possible. He offered to meet up with me the next day and show me around, and when we reconvened, he told me of his plans to work in Italy upon graduation as he sped me through the sights. We practically sprinted through the local museum, with its old farm implements, dishes, commemorative certificates, and pictures of local history. Even the military memorabilia of the guerrilla fighter Michael Collins, who survived the war of independence against England but not the factional squabbling that came afterwards, couldn't diminish his pace. He slowed down outside to speak with tender enthusiasm of the details of football culture in Ireland, of how the whole country organized itself around live World Cup broadcasts, and of the impending World Cup. When I asked him whether he and his peers followed politics at all, he shrugged and said, Football politics. Like many younger people I met, he spoke of the conflict in the north more as a brutal mess than a meaningful struggle and saw nationalism and national history as another generation's passion.

The four older people I had dinner with in a little house in the heart of the city still regarded the war of liberation as unfinished and themselves as parties to it. We're still fighting, they said over a table crowded with curried eggs and pastas and salads and water goblets of Waterford crystal followed later by whiskey and coffee, and at midnight we were still talking. I had met the archeologists Lee and Paddy seven years before. She was from Colorado, a cousin of the man I'd come to Ireland with on that first visit, and had been settled with Paddy in southwestern County Cork for many years. In a country without divorce, however, they could never marry, and so Lee was still a foreigner and Paddy still had a wife around somewhere (divorce finally became legal in 1996). Magnificent with her long graying hair piled up on top of her head and her straight back, Lee would look good driving a chariot. Paddy was mild and scholarly with his white beard and blue eyes, fairskinned like his sister Mary and her husband Dennis, the hosts of our feast.

All through the meal and into the night, they told me stories, and the conversation rolled forward in a series of anecdotes that brought forth other anecdotes. American conversations tend to be dialectical, a give and take of short statements, or even more laconic, the kind of monosyllabic exchanges so popular in tough-guy texts and television. In the West and even more particularly in the Western, silence is a sign of strength. Ireland has a different conversational economy, one in which the ability to talk well is a gift and perhaps even a weapon, for the political disenfranchisement and powerlessness of the Irish people and the Irish language under an English government are often described as silence. Someone once suggested to me that styles of speech resemble the landscapes they emerge from, that one can trace the flatness of the plains in raw midwestern accents, desert silence in western taciturnity, the lushness of the southeast in its denizens' dulcet tones. The same could be said of Ireland, whose intricate, winding landscape is so densely and intricately scored with the stones and wounds of history, and whose musically rising and falling speech can hardly proceed without anecdote.

At our dinner, words accumulated and meandered, drawn onward by the magnetic fields of memory and association. Everyone sat leisurely while a story was told, then brought forth another story, personal or comic, as a response. Paddy began one by explaining to me that Knock means hill, and it's also the name of the town in the north where the Virgin miraculously appeared, not in the flesh but as an apparition, in the last century. A man is crossing the border back into Northern Ireland from Knock, and the border guard says, What's in that bottle?, and the pilgrim says, Holy water, and the guard opens it and says, It smells like poteen (bootleg whiskey) and the man cries, A Miracle! His sister Mary told a related story about her youngest child's first communion, and then they began to tell immigrant stories, since I was something of one myself.

Their grandfather from the west of Cork had been sent the money by his emigrant older brothers and sisters to join them, and he hopped a boat – in those days, they said, you didn't need passports or visas – got through Ellis Island and New York to Boston, where the family was, took one look at the place and came straight back. When he was old and losing his memory, he would say several times a day, I knew New York like the back of my hand. It reminded Dennis of a story about the woman who took care of his grandfather and didn't know how many brothers and sisters she had and, because she was illiterate, lost track of her son in America. This too turned into a story: The post office gets a letter addressed to My Son in America and delivering it becomes a challenge that drags on and on – until one day a man walks in to mail a letter addressed to My Mother in Ireland, and they know they have their man. The others talked about how the families who remained behind wondered about those who emigrated, that even those who weren't uprooted felt a loss I had never heard about, the loss of the left behind.

Only at one point did the storytelling rhythm of the evening dissipate: when Mary and Dennis began to tell us, in tandem, about a television program on cheetahs they'd just seen. They spoke of it not with the ready gape at superlatives and statistics sheer entertainment provides, but with a tender awe at what biology was capable of producing. There was a note of yearning in their voices for such marvels, for the bursts of speed and power the cheetahs represent in an undiminished primal landscape, or for the real presence of the totems and symbols that linger in the imagination. Sometimes the wolf at the door should be a real wolf, but Paddy said the last wolf in Ireland had been killed in 1792, on Mount Gabriel in the far southwest of Cork. I told them about the woman who'd been killed in California, not long before I left, by a mountain lion, and about the black bears that had returned to the mountain in the center of the county I grew up in, about the ways in which my own landscape was still marvelously wild, even if almost everyone in it was too inarticulate to tell a story.

Around midnight, after we had all emptied our first glass of whiskey, the son who had figured as a naive child in Mary's stories came home, a man of my age. They insisted he drive me to my lodgings, though I was unintimidated by whatever the nights of Cork could hold. On the way there, they instructed him to take me on a detour, to St Ann's Church, Shandon, across the river in the heights of the city, so that I could see the golden salmon atop the spire of the church, the highest thing in the city, and I saw it, swimming in the night sky. Whatever the fate of the live animals of Ireland, their images circulate still – literally, too: the coins, which Yeats designed, bear salmon, stags, horses, a sort of kingfisher, and, on the small change, a couple of intricate mythological birds.

It was a story Lee and Paddy had told me seven years ago that impressed upon me how different the sense of time here could be and that had drawn me back. We had come to stay with them straight from the airport, and so we were still slightly dazed by jetlag when they took us on an outing a day or two afterward. They led a local archeology club in west Cork, the Mizen Field Club, which once a month went on an excursion in a rented bus. Its members were mostly local farm families, though I remember a defrocked minister and a stroke-silenced local historian among the people who embarked with us on that drizzly Sunday in May. The bus wound us over a steep landscape in which every third house seemed ruined and abandoned, to the Mizen Peninsula, the second of four peninsulas sticking out like limbs or fronds on Ireland's southwest coast. The sites we visited were determined more by proximity than by chronology, and for the locals they must have all fit into a general picture. But for us, it was a jumble, as we shot from pre-Christian Celtic sites to those of Elizabethans and the IRA along the coast. And at each stop in the steep coastal landscape Paddy got up before us and recited the events that had unfolded on the ground where we stood. The idea of songlines in Australia – long trajectories across the landscape navigated by narratives of the Dreamtime – has become familiar, but the Irish countryside Paddy guided us through seemed more a palimpsest than anything so neat as a directional map, scored and scribbled over by thousands of years of events running in every direction, almost all of them marked by some further rearrangement of the superabundant local stone.

What I remember best is the great stone pile of a fortress where O'Sullivan Beare's followers had heroically and unsuccessfully resisted an English siege in 1603, one of the last great battles before the Flight of the Earls, which marked the end of the indigenous aristocracy; and a few miles away another ivy-tangled ruin, of an Anglo-Irish stately home destroyed by the IRA in the 1920s. Better than these reciprocal ruins even, I recall a stone that had only been altered by the addition of legend. A rough, hulking, henshaped thing at the top of a steep slope above an inlet of the sea, it was said to be the Hag of Beara, the wife of the god of the sea who had been turned to stone by a resentful priest and left to brood forever over the nearby coast; she appears later, in a ninth-century poem, as a withered mortal recalling lost beauties and lovers. On the way back, the little girls sitting together in the back of the bus amused themselves by singing, in thin, good voices, a long gory ballad whose refrain went, "They hung 'er by a ribbon from the sour apple tree/And we won't see her no more."

Afterwards Lee and Paddy took us to a tiny pub in a town that was little more than a place where the road was paved and lined with houses. The walls of the room we sat in were lined with corrective horseshoes of wildly varying size and shape, labeled with all the equine afflictions they were made to address. It was at the end of that day, in the dim, diminutive pub, that I heard the story, which they had heard themselves from their near neighbor.

He was a nonagenarian and a native of the area, and when he was a small child near Skibbereen at the turn of the century, a beggar would come round every two or three months, quite regularly. His mother would always give the beggar a meal, but the tale-teller himself would always hide in the lane when the beggar came, frightened and fascinated by his walk, a walk for which the other boys mocked him. (Somehow I came by a mental picture of the whole scene: a pale dusty road lined with hedges and the view from around a hedge of the beggar coming up it.) The man scuttled with a crablike gait, legs turned out wide and knees bent.

One time his mother told him the beggar's story, to explain the gait and dissipate the fear. The southwestern counties of Cork and Kerry had been worst hit by the potato blight that struck in 1846 and returned several times in the following years, and so many died at the height of the Famine that funeral ceremonies were abandoned. Instead two men with a cart were sent around each day to collect the corpses and bury them in a pit. The children of one poor couple had died, all but one small boy, and the mother couldn't bear the thought of throwing this last child in the common grave. She prevailed upon her husband to make a coffin from the few scraps of wood they had, and the coffin he built was short but wide. They had to break the dead child's legs to make him fit, and then they gave the coffin to the carters to take away. Returning later that day with another load of corpses, the carters heard a mewing sound from the pit. It was the boy in the coffin, not dead after all, and it was he who eventually became the beggar who used to circulate around County Cork.

It was an appalling story, but for me it was also a marvelous story. I had thought of the Famine as something irrecoverably distant, as far beyond the reach of living memory as the Inquisition or the sack of Rome, for my own history at that time had hardly included tales that stretched back before the Second World War. That there was a man living who remembered a survivor of the Famine – that the link to events of a century and a half ago was unbroken – was a discovery of astonishing delight. Time itself is elastic: the past is kept breathing by speech or buried in silences. About a hundred and fifty years seems to be about the farthest reach of living memory, the length of time encompassed by an old person who as a child encountered a survivor of some long-ago drama. The story of the beggar was my first intimation of how far back memory could stretch.

Picture what is remembered as a kind of daylight shared by the living, and picture its farthest reaches as the place where twilight is falling, beyond which is only the blackness of oblivion and the dryness of recorded history. That the day before the dark could be so long I learned for the first time over a glass of beer in that dim room full of corrective horseshoes. For historians the reach of living memory makes an immense difference in the nature of their work and forms the boundary to a certain kind of history, one yet unrecorded, still surviving in conversation, a history that does not yet belong entirely to the past but is still alive in the present. Time itself is elastic, uncertain, the past brought close by speech or buried in silences. There are moments of passage, a kind of amber late-afternoon light, in which events assume a final determining significance before those who remember them begin to fade away, and the event becomes nothing but history; one can watch all the atrocities of the Second World War now, with their long shadows stretched across us still, retreat from the personal to the public, from memory to history, as the 1919–22 Irish War of Independence largely has (though Paddy still had stories to recount which he'd heard from participants).

It is fortunate that so much work has been done already on such events as that war, for those passages of people's lives are entering the night of the grave and the book. There have been moments and figures that historians have cared about too late, so that there are few firsthand accounts with which to work, and the evidence is all indirect; nothing can make it come to life again. The Famine which determined the crippled beggar's life is something like that. My hosts' old neighbor remarked that no one would talk about it when he was growing up in Skibbereen, and most accounts of the Famine come from horrified witnesses and rationalizing administrators rather than those who experienced its effects firsthand. Silence is one of the elements of the Famine that witnesses describe over and over again, the silence of those who died and those who were too weak to separate themselves from the dead next to them, and the silence of those who toiled like wraiths on the road-building and stonebreaking relief projects by which they earned a bare survival. A nineteenth-century archeologist who tried to recover that history wrote, "This awful, unwonted silence, which during the Famine and subsequent years almost everywhere prevailed, struck more fearfully on their imaginations, as many Irish gentlemen informed me, and gave them a deeper feeling of the desolation with which the country had been visited, than any other circumstance. . . ." Trauma is inherited as silence, a silence it may take generations to learn to hear.

I grew up in a place without a past, the shadeless artifice of suburbia. My parents hardly spoke of their own past, and no one had anything to say of the past of the place we'd landed in. My father told me exactly three stories about his childhood in an immigrant ghetto in East L.A., each overshadowed by public violence such as the pogroms, and there were seldom relatives around to make up the lack. The real function of my mother's tales always seemed to be to demonstrate that she and her brother and sisters had been sunny paragons by comparison with her own monstrous brood of boys and me (and they probably had been, though Uncle Dave once teased his mother, when she made the same point about his five children, We weren't good, just malnourished). We lived in a house in a new subdivision where everyone else seemed new to the area too, except the old couple in the big square two-story farmhouse across from the beginning of the street. They must have sold the farmland that became our tract, but they still kept the hill around their home rural, with a grumpy old sheep, several Herefords, ruinous barns, and a working windmill arrayed around it. Their house was just off a road that began as a main street in town and faded into a dirt track that came to an end in a horse pasture. The last street off this road was ours, a double strip of new houses all built according to two or three sets of plans, but painted and trimmed to suggest individuality. The neighbors were mostly conservative white people whose unimaginative offspring chose to play in the street far more often than on the hills up the road. Even the plants around the houses were baleful things with blue needles, inedible fruit, glossy evergreen leaves, as much like plastic as any living thing could be, nameless plants that had no symbology, no meaning, and no use, but for their low-maintenance decorative qualities. The pieces of this subdivision didn't yet add up to culture, something built out of stories about which further stories can be told.

The uneventfulness of the Golden Age is storyless, best celebrated in lyric poetry and static paintings. Story, and history, begin when something goes wrong, when a child appears to die, when a woman talks to a snake, when a man is hung up upon a tree. But to be changeless and storyless in anything less than the Golden Age is to be stuck in tedium. Stephen Dedalus said in a famous line of *Ulysses,* "History is a nightmare from which I am trying to awake," and contemporary Irish writers complain about the American appetite for the Irish past; the Caribbean writer Derek Walcott penned an implicit reply: "Amnesia is the true history of the new world." History hovered outside the borders of my own sedative neighborhood like one of those dreams that are all the more alluring for being just beyond recollection. That the primary subject of that dream was suffering didn't dissipate its enchantment for me, and still hasn't. I didn't hear histories until later, only read them in books which I piled up around me like the stones of a fortress, against not only a family and a neighborhood, but even an era that seemed unsympathetic and blankly amnesiac. Now a number of long memories have come my way as stories, of the Indian wars, of southern families, even a few of my own family in Ireland and in the Russian-Polish Pale, but no place seems as haunted and infested by stories, and by the past, as Ireland.

I heard a story once about a tourist who goes to a pub in Ireland, where the locals so incite him with outrage over the deeds of the English that he starts off drunkenly to right the wrongs, and they have to hold him down and explain that it all happened several lifetimes ago, however vividly it lives in their conversation. I met an Irishman who'd emigrated to South Africa and cursed Ireland for its inability to outlive the past: his new home, he asserted in the wake of President Mandela's inauguration, would set the past behind it, heal the historical wounds, and invent a new nation. At least for outsiders much of Ireland's charm is that it is still, however literate, an oral culture. Talk is a principal form of entertainment and an art, and internal memory hasn't been entirely eclipsed by recorded history or amnesia. Storytelling itself has been in a long decline elsewhere, in part because the generations are all but segregated in most industrial societies, because a tale requires a leisurely pace for both teller and listeners, and because telling has been replaced by commercial entertainment. The appetite for stories seems undiminished, but the information and entertainment media have evolved to fill it with narratives in which the listener is forever inaudible and invisible, never the teller or part of the tale. These sources don't really replace firsthand stories, which cast their glow over the events and places of one's own life, incorporate one into a community of meanings.

There are other measures than the bucket brigade of memory by which to gauge the repercussions of an event. Those of the Irish Famine are still being felt. It was the fulcrum which changed Catholic Ireland from a nation of entrenched peasants to a wellhead of emigrants, and that torrent which populated and helped shape much of the rest of the English-speaking world is still flowing. The immediate source of the Famine was the potato blight, which caused the crop suddenly to rot in the fields or in storage, with a terrible stench, partially in 1845 and then almost totally in 1846 and 1847, and to varying degrees for years afterward. Much of the impoverished majority lived close to the edge under ordinary circumstances, and numerous previous famines on a smaller scale preceded the several years of potato blight that would become known as the Great Famine. It affected potato crops across Europe, but only in Ireland was the population so dependent on this one vegetable that the blight brought disaster. The scope and horror of the disaster – of widespread starvation, the intermediate malnutrition which brought on blindness and insanity, and the concomitant spread of diseases such as typhus – was incomparable to anything of its time; historians still reach back to the Black Death, the fourteenth-century plague that devastated all of Europe, for a parallel.

It is fair to say and often said that though the Famine was in its origin a natural catastrophe like the Black Death, its effects were drastically magnified by politics and economics. Had the Irish poor and their spokespeople been less inaudible in the English Parliament, had the late and scanty famine relief allocated by England not required many small landholders to give up their land to receive aid, had the potato growers not already lived in a poverty that made them utterly dependent on a single crop, without cash or commodities to exchange for other food, the outcome might have been far different. As it was, Ireland remained a food-exporting country throughout the Famine – three-quarters of its farmland was planted to grain inaccessible to the majority of its potato-eating population. So they starved amid plenty. The Famine was itself, like many contemporary ones, a consequence of the distribution of wealth and power rather than of the fickleness of nature or absolute scarcity.

The Famine changed the face of Ireland forever. In affecting the poorest of the south and west most drastically and causing them to emigrate or die in greatest numbers, it accelerated the loss of the native language and culture which had been begun by the education act of 1831, legislating universal education – in English, as English citizens. So it was also a catastrophe of forgetting, and in a recent spoken-word song, the rock star Sinéad O'Connor conflated the education act and the Famine into one traumatic silence:

> And so we lost our history,
> And this I think is what's still hurting.

And she steals back a little something from England, the refrain to the Beatles's "All the Lonely People." The Famine rapidly shifted a great deal of land from very small farmers to much larger ones and began the elimination of the rural community that seems to be reaching its final stages now; it is sometimes said to have marked the end of the Stone Age in Ireland by ending some of the primordial way of life and agriculture that came out of rural poverty; and it pushed a nation of people who were exceptionally attached to their homes into ships on one-way tickets.

The historian Kerby Miller concludes that before the Famine the Catholic and Gaelic-speaking Irish were, unlike the Scots Irish who had already mobilized for America in disproportionate numbers, exceptionally reluctant to emigrate. Among the reasons he enumerates are: their profound attachment to place; a language in which the word for going abroad translates as *exile*; a literature in which all departures from one's country of origin were regarded as tragic exile; and a syntax that encouraged and reflected passivity and fatalism. A culture that deemphasized individualism, initiative, and innovation and discouraged breaks with tradition and community was one in which departure would inevitably be a forced exile rather than an embraced opportunity. Only such a catastrophe as the Famine could loose the ties of place and tradition and propel the population outward. The rapid drop in the population owes more to this than to death itself. About a million are thought to have died of hunger, exposure, and disease resulting from the Famine – and during the Famine years, a quarter of the population, more than two million people, emigrated, mostly to North America.

Emigration seems to Ireland what the western frontier was to North America, an outlet for the unemployed, the restless, the insurrectionary, and for a swelling population. Ireland is almost unique among the nations of the world in having a population that declined during the past century and a half; the North and the Republic together have not much more than half the population of 1846. One interpretation suggests the real trauma, the subject of the central silence, was emigration itself. The historian Joseph Lee writes, "The second major distinctive experience of modern Irish history, next to colonization, was emigration. Emigration was not unique to Ireland. But the type of emigration and the impact of emigration was. In no other European country did emigration become a prerequisite for the preservation of the social structure and the status system. . . . After the Great Famine, the emigration of 'surplus' children became a necessity for the farm family if the parents were to transmit the inheritance intact, and for the labourer's children if they were to survive. But emigration came to be perceived as a blot on the communal landscape, a shaming indictment of the incapacity of the nation to provide for its own people, just at the time when a national consciousness was beginning to spread through the populace. . . . No other society found itself obliged so remorselessly to rationalize the subversion of the family ideal inherent in the emigration 'solution.' . . . The psychic impact of emigration, the price paid both by those who stayed and those who went, for the subterfuges and the evasions to which the society had to resort to preserve its self-respect while scattering its children, has scarcely begun to be explored."

During the brief period of prosperity created by industrialization and the expansive world economy of the 1960s and 1970s, it seemed that emigration as a recurrent feature of Irish life was over. But in the 1980s the new urbanized, postagricultural economy began to crumble, and while family farming continued its long decline, the city jobs that had absorbed rural refugees dried up. The high Irish birthrate has long demanded either industrialization or emigration as an outlet for the young pouring into the job market, and industrialization has not been a great success despite all the tax and pollution breaks offered foreign and multinational corporations. A new wave of emigration began and continues. In the 1980s one person out of twelve in the Irish Republic emigrated, the majority to England and the States; in 1994 alone more than seventeen thousand people from the Republic were allowed to legally emigrate to the US (and despite the disproportionate number of visas granted since the 1990 Morrison Act, the US also has a vast, but uncounted, population of undocumented Irish immigrants).

Counseling and aid with emigration are now standard offerings of church and community youth centers, but the conservatism of the country is a factor in the flight of the young, who often give the impression of running away to join the twentieth century and embrace the rootless and urban with satisfaction, if not without backward glances. At one point on my journey, I met a woman in her early twenties who had gone back to her village for a festival and found that everyone she had gone to school with, everyone her age, had emigrated outright. She was the only one of her peers still in the country, she told me, let alone the village. Versions of her story are common, and it seems sometimes that Swift's *Modest Proposal*, in which he proposed that the impoverished adults of Ireland eat their young or sell them for consumption, has come true by other means. The century and a half of emigration has consumed the population, particularly the young, as even the Famine didn't; and unlike those who died disastrously at home, those who went abroad to work became a commodity – they are still cheap labor in their new countries and were formerly a resource in the form of money sent back to the old one. When he visited Ireland, John F. Kennedy said, in words that must have been meant to be charming but are instead chilling, "Most countries send out oil, iron, steel or gold, some others crops, but Ireland has only one export and that is people."

The beggar in the story wasn't an emigrant but a migrant, and the image of him rotating around southern Ireland as regularly as the hands of a clock conveyed something of how established begging was when charity was still considered a Christian duty. Foreign writers of the eighteenth and nineteenth centuries routinely complain about the quantity of beggars, but their continuing presence suggests that begging must have had some kind of niche in the society. For centuries Ireland had a huge population of wandering beggars, both the permanently homeless and the wives and children of impoverished peasants who would take to the road during the mild months while the potato crop was growing and bring back money to help pay the rent. The Famine displaced vast numbers of families who had once been tied to the land, because famine relief was unavailable to those who owned any land at all and the failure of the crops forced others to sell their land or lose their leases. The population of beggars took a temporary leap upward, and millions took to the roads; so that in addition to the outright emigrants a new swarm of migrants arose. Nowadays begging is a last resort, and beggars are considered to have dropped out of the world with its schedules and routines, but begging was once a living or at least a way of life.

Lee's and Paddy's beggar was himself a conundrum, this man who had been crippled by love and saved by strangers, who been resurrected in the midst of terrible death, an outcast who had become a fixture on the rounds of his route, homeless and local, a lame man whose principal activity was walking this circuit over the seasons and years, in one year and out the other as liberation movements and dreams rose and fell, millions emigrated, the world lurched forward into a century of wild growth, destruction and unimaginable transformation.

Lee and Paddy insisted I stay with them for a few days after the dinner in Cork. They lived in an old farmhouse outside the hamlet of Ballydehob, whose name has become the eponym for rustic and remote living, the Dogpatch or Hicksville of Irish jokes. I never met their old neighbor, now ninety-seven, who recalled the beggar – I don't think he was ready for anything new at his stage of life – but I did hear him. Having finally come to terms with the possibility he wouldn't be around to tell them forever, he had committed some of his stories to tape, and they played the tapes for me. It was clear he was steeped in the past that preceded his own birth, for he spoke familiarly of the Great Snow of '55 – 1855 – as the standard by which all other cold spells were judged, and spoke, the way one might of the recently deceased, of a local poet who had gone to the American Civil War as a reporter and died two days before Gettysburg. He told a slightly different version of the story of the beggar with the peculiar gait, though it only included a few more details than my hosts had recalled seven years before. The beggar's name was Tom Guerin and he was, like so many in this island, a poet. In one of his poems he portrayed himself as a man with one leg pointed east and one pointed west.

Lee and Paddy were great hosts whose conversation was a gift, who stuffed me several times a day with gallons of tea and local delicacies, and who couldn't travel across their home territory without a stop to show me a carved cross or a ring of standing stones tucked away inconspicuously in a cowfield. Probably tired out by their own generosity, they packed me off on the last day of my visit with a snack and some vague directions to go see some petroglyphs a mile down the road and a ring fort a mile or so up the mountain in the other direction. I put my inch-to-two-mile map of the region in my pocket and set out for the fort on the side of Mount Kid on another May day whose clouds were sodden and distended, though they seldom brought forth much rain. The eighth-century stone cross they had shown me in a cowfield near Bantry had a vertical boat carved on it, sailing straight up to heaven, and the highest thing in Cork was the golden salmon atop the church; sea and sky seemed easy to confuse here, with the earth nothing but a slice of stone and melting mud between wet clouds and hungry sea.

It was easy enough to find the dirt track I was to turn off on. Not much further down the road, a flock of belligerent sheep raced up to the fence as though they would trample me if they could and bleated with the sound of a traffic jam in Arcady, and dogs lurched out at me the rest of the way, so I walked with handsful of rocks. The ring fort was nothing more than a circle of low stone walls whose cow pies demonstrated its most recent use, but it was a vestige of the petty warfare of early Christian times, another one of the obdurate witnesses to all the generations who had crisscrossed the landscape before. There were ghosts, quiet ones.

My next instruction was to admire the view from the top of Mount Kid, a hill about a thousand feet high on the spinelike ridge stretching up from the Mizen Peninsula. One of the peculiar features of this terrain is that the highlands are often the boggiest; the deep soil has become impermeable, water no longer drains downward, and the peat above acts as a sponge to hold it in place. Mount Kid's west face became increasingly squishy until I was hopping laboriously from tuft to tuft of the moss and sedge that grow like tiny islands on the wet ground. Along the flat back of the mountaintop was a square pit where someone had been cutting peat. An accumulation of plantlife whose wetness has kept it from fully decaying into soil, peat or turf is dried to become one of the principal fuels of Ireland – even some of its power plants are peat-fueled – and one of its characteristic sights and smells: the stacks of turfs piled near rural houses, the sweet smell like molasses and grass of a turf fire. (When Friedrich Engels attempted a history of Ireland, he started with the bogs and the lack of better fuel. "It is obvious that Ireland's misfortune is of ancient origin," he wrote. "It begins directly after the carboniferous strata were deposited. A country whose coal deposits are eroded, placed near a larger country rich in coal, is condemned by nature to remain for a long time the farming country for the latter when the latter is industrialized. That sentence, pronounced millions of years ago, was carried out in this century. We shall see later, moreover, how the English assisted nature by crushing almost every seed of Irish industry as soon as it appeared.") The bogs themselves are considered threatened in many places, by reforestation, burning, and peatcutting; and no intact large bog remains, only the small bogs such as the soggy patch of mountain I was skipping across.

As I ascended, the peninsula became visible, crowned by Mount Gabriel, where the last wolf was said to have been killed. The islands scattered in Roaringwater Bay to the south were a pure deep blue, and the change from the green underfoot to the blue of distance was perceptible in all its phases. The land between the mountain and the islands was agricultural, a collection of fields marked off by the wavering lines of roads, and to the north were the higher ridges of the mountains between Cork and Kerry, also that exquisite wistful blue that painters relied upon to indicate distance in the days before vanishing-point perspective expressed the transformation from here to there. Such a blue is deeply satisfying by itself, but it is also pleasant to contemplate as visible evidence of the materiality of the atmosphere.

Tired of the slow process of locating dry patches in the bog, I descended on the rocky eastern face and found, where the slope began to flatten, a ruined farmhouse with a warren of outbuildings around it, set amid fields so stony their outcroppings often towered over the stone walls. A pair of pheasants burst out of one such clump with a booming cry, alarming amid the quiet. We had explored a ruined house on the side of the mountain several years ago, whose downstairs was dominated by what had once been a blue-painted room with a cheap iron fireplace. Green mold had grown over the pale blue plaster and given it a sad and elegant iridescence; the two rooms upstairs had been painted pink. But this house was far more ruined; its floor an impassable heap of sharp shards of wood, its walls crumbled back almost to the stone, and its staircase nearly gone. Only the stones of the house would remain in a few years more, though the squat stone outbuildings that must have been for peat and chickens and the like were still standing. Suddenly I saw a square window a few feet from the rough ground in one outbuilding and remembered it, not from my visit but from the photograph my lover, Lee's cousin, had taken of my head framed in the wall. I was back at the same place after all, though we had both changed, the house and I. I had a strange compulsion to kneel and replace my head where it had hovered seven years before, but I knew that to do so would be a terrible mistake.

By that time it was not the ghosts of former inhabitants and ages past haunting me, nor that of my lover, with whom I had separated a year later when it became undeniable that our lives were taking different directions, but the ghost of the woman I was then. Between Lee's recollections of his family and mine of the photographs we had taken, the past had come swarming back and brought with it a sense of terrible loss. I had to recite to myself all the fine things that had happened since to ward off that blue melancholy, but at the window, I suddenly felt my own former self dead here, with her dreams and all her plans for futures that never came to pass. Every cell in the human body is renewed every seven years and so nothing material remained of the younger, more timorous woman who had come here before me, and nothing but a bridge of memory connected us, memory so frail it hung on a vacation photograph.

6 Anchor in the Road

From the edge of Roaringwater Bay in Ballydehob, I set off walking around noon one day for the town of Bantry, finally alone and afoot as I had planned, or almost as I had. Before I'd left, I had sat down with a map of Ireland and made an itinerary for walking from one small town to the next along the west coast, and the list of names themselves were an incantation, Bantry and Kenmare and Killarney and Tralee and Listowel and Glin and on and on from south to north. Anticipation itself is a great pleasure, and planning is a fine way to bask in it; I have enjoyed making lists for travels since I plotted to run away from home with my middle brother when I was eight or nine, a plan that never got beyond the list of survival gear I thought we should take with us to the hills. My Irish itinerary fell by the wayside too.

In Dublin I had found a mountaineering store and bought a clutch of detailed topographical maps from a shaggy, lithe boy with long sideburns, who talked with me in a gentle murmur in which the accolade *grand* shone like a minnow again and again, silvery with pleasure. Grand were my plans to walk up the west, was the surfing in Clare where he was going to catch some waves during the bank holiday coming in a few weeks, were my stories about the marvelously inventive language of surfers in my own part of the world and the Cliffs of Moher I should go see in Clare, his home county. The maps he sold me, pretty squarish redcovered things with sea level in green and altitude shading into the soft orange of 2000 feet, with a name on every square mile at least of land, made it clear that the route I'd daydreamed over was impossible. They showed that the roads were laid out less as the crow flies than as the butterfly does, often doubling the distances shown by the less-detailed maps.

On my previous trip my sense of scale had veered wildly, shaped as it was by continental and interstate scale: I would look at a place on the map and say, We can't go there, it's all the way across the country, then realize the country was hardly a hundred and fifty miles across, only a few hours on a straight highway, and finally find that the slow winding roads of the countryside keep the island large after all, in a way that has nothing to do with notions of objective scale. The roads had been built not for long-distance travel but to to connect the dots of adjoining towns, and they do so in serpentine lines that writhe even more to accommodate the steep terrain of the west. Still I planned to walk, but it was already becoming clear that to do so continuously along an extended contiguous stretch of land was not nearly as alluring as it looked on a map.

But traveling, being in motion, is itself profoundly satisfactory. Most stories are travel stories, and in traveling our lives begin to assume the shape of a story. It may be because a journey is so often a metaphor for life itself that journeying is satisfying. In motion it seems that time is not slipping away from us but we are pursuing it, measuring its passage in the rhythm of the road, the metaphor become literal. Perhaps if we didn't imagine life as a journey rather than some other metamorphosis – the growth of a tree, for example – roads would not seem like destiny itself, but we do and they do. To move along the road is to encounter all the loose elements, the dangers and possibilities, to slip out of a settled destiny in pursuit of stranger fates. The road is a promise as simple as what lies ahead, never failed and never delivered, and the road is a strange country itself, longer than all the continents and narrow as a house, with its own citizens, its own rules, a place where the solid and settled become fluid.

There are domestic dramas in which all the turmoil passes within the walls of a home, but the romance is the central literary form of Europe, one that may have reached its apotheosis in America: the romance as quest, adventure, pilgrimage, crime spree, road trip, travelogue. And however celebrated the notion of home, however fixed an idea of the Irish as a sedentary people uprooted only by crisis, Irish literature seems particularly immersed in the charms and potentials of the journey. The ancient queens, kings, and heroes wander from host to host and from battle to home; the saints are mostly voyaging saints; the Gaelic-speaking poets who composed much of the literature before the nineteenth century were themselves wanderers; displacement and exile became principal poetic themes; and the centerpiece of twentieth-century literature, *Ulysses*, regenerates the travels of Odysseus in the details of a Wandering Jew's day in Dublin, compressing the resonances of travel and exile into the compass of that city. J. M. Synge celebrated the road and its turn-of-the-century population of tramps, vagrants, balladsingers, and flowerwomen in his plays and essays and often roamed himself. "Man is naturally a nomad . . . and all wanderers have finer intellectual and physical perceptions than men who are condemned to local habitations," Synge wrote in his notebook, adding "But the vagrant, I think, along with perhaps the sailor, has preserved the dignity of motion with its whole sensation of strange colors in the clouds and of strange passages with voices that whisper in the dark and still stranger inns and lodgings, affections and lonely songs that rest for a whole life time with the perfume of spring evenings or the first autumnal smoulder of the leaves."

There was a tale, "The Destruction of Da Derga's Hostel," that had stuck in my mind from the time when I yearned for a prehistory of my own and read lots of Celtic myths; it was one of those balladlike tales weaving together wandering, the supernatural, bloodshed, doom, and a skein of place names. An impacted myth, my translation called it, as though it had been crushed together by a collision of disparate elements, or as though it was jammed below some literary gumline like a wisdom tooth. An ancient story when it was written down in the twelfth century, it begins with the famous beauty Etain's daughter, who was abandoned, raised by cowherds, and then chosen by a barren king to be his bride. The night before she was to be taken, "she saw a bird coming to her through the skylight; it left its feather hood in the middle of the house and took her and said, 'The king's people are coming to destroy this house and take you to him by force. But you will be with child by me and will bear a son, and his name will be Conare and he is not to kill birds.'" When the king died, his men prepared a bull feast, a method of divination whereby a bull was slaughtered and a man "ate his fill and drank its broth and slept, and an incantation of truth was chanted over him. Whoever this man saw in his sleep became king. . . ."

While the bull-dreamer was dreaming, the young Conare was playing with his foster brothers on the Liffey. There he saw huge white-speckled birds and chased them until his horses grew tired, then pursued them on foot until he reached the ocean. They flew over the waves but he overtook them at last. "The birds left their feather hoods, then, and turned on him with spears and swords; one bird protected him however, saying 'I am Nemglan, king of your father's bird troop. You are forbidden to cast at birds, for, by reason of birth, every bird here is natural to you. . . . Go to Temuir tonight, for . . . the man who naked comes along the road to Temuir at daybreak with a stone in his sling, it is he who will be king."

Temuir was Tara, the royal hill not far northwest of Dublin, and of course Conare went there, naked as he had stripped to hunt his kin the birds, and became king. He was bound with a great number of taboos and, in the protection of his foster brothers from justice, he managed to break them all. War broke out, he became the king "whom the spectres exiled," and the fighting closed off all but one road to them, the road to Da Derga's Hostel. Da Derga was a red god of the underworld, and one taboo forbade Conare to follow three red men to this god's house. But even the teeth of the men he and his retinue followed were red, and at the house with seven doorways and a road and a river passing through it, Conare met his fate in a siege. One of his followers who had left to find water to sate Conare's thirst came back to find him decapitated, and poured the water straight down the throat of the headless torso, while the head recited a few lines in praise of his thoughtfulness. Severed heads, like birds, talk a lot in Celtic myths. But it was that moment in which Conare was walking naked along the road that spoke of the power of roads, those places where one might be anything and nothing. Roads are a no-man's-land, a leveling ground, the place where one is no longer one thing and not yet another, the paradoxical place the crippled beggar from the Famine occupied as a home. There's a nearly straight road from Dublin to Tara now, and a golf course when you arrive, in case the contemplation of ancient majesty is not sufficient. Perhaps the eighteen holes of a round of golf signify a journey too, a secular Stations of the Cross, but not for me.

I was travelling from Ballydehob to Bantry, and I crossed the bridge over the Bawnaknockane River where it enters Roaringwater Bay, passed Knockroe, Knockaphukeen, Coosane, Ballybane, Barnaghgeehy, Letterlicky Bridge across the Durrus, Hollyhill, Cappanaloha, to Bantry Bay – a fine collection of names for a ten-mile walk over a range of modest hills. The road had no shoulder, so that what was most of the time a sedate pastoral became a racetrack whenever a car whizzed by. It was lined with low stone walls and hedgerows full of flowers profoundly familiar and subtly different in equal measure: owl clover, buttercups, violets, ferns, bluebells, fuchsias (an introduced species gone wild), gorse, the occasional tall foxglove nodding its specklethroated bells, and tiny pink and blue flowers I did not know. Cows bellowed, sheep bleated, birds chirped, and occasionally cars roared, all under a sky that seemed distended and sagging with its burden of water, though it only dripped a tear or two here and there. By the time I was almost in Bantry the clouds had grown thinner and then worn out like rags and the sky shone through in patches. My shadow suddenly leapt into existence ahead of me on the road, as I passed an anchor about a mile before town, an anchor like a huge rusted cross with a rocking base. A plaque said it had been fished out of Bantry Bay in 1964, a relic of the French fleet's unsuccessful invasion in 1796, instigated by Theobald Wolfe Tone.

Son of a Dublin carriagemaker, Wolfe Tone had been an idle and charming young man galvanized by the American and French Revolutions. He founded the nondenominational United Irishmen (and United Irishwomen) in Belfast in 1791, in that spirit of universal brotherhood which had stirred the revolutionaries of the time, and made it too into a revolutionary instrument. Like a great many Irish patriots, his love of justice and homeland exiled him, and he spent much of the last years of his life marshaling support from abroad. The invading French fleet he recruited in 1796 had the worst possible weather: calm when they needed wind to cross the sea from Brittany to Bantry, foggy when they needed vision and clear when they needed stealth, and finally, wrote Wolfe Tone that Christmas Day, "It blew a heavy gale from the eastward with snow, so that the mountains are covered this morning, which will render our bivouacs extremely amusing. It is to be observed, that of the thirty-two points of the compass, the east is precisely the most unfavourable to us." Having lost their element of surprise and some of their ships, the French gave up and sailed out of Bantry Bay without a battle. The invading fleet was made up of forty-three ships whose names convey something of the optimistic momentum the French Republic still possessed: Indomptable, Nestor, Droits de l'Homme, Patriote, Pluton, Constitution, Révolution, "and the unfortunate Séduisant," Immortalité, Impatiente, Tartare, Fraternité, Fidélité, Atalante, Justine. . . . Captured by the British and sentenced to be hanged as a common criminal rather than shot as a rebel, Wolfe Tone slit his own throat in a Dublin prison in 1798.

It isn't Wolfe Tone and the events of 1798, but St Brendan and the legends of the sixth century that have pride of place in Bantry, a quiet little town whose streets all ran down the hillsides to converge near the harbor, where a big modern statue of St Brendan the Navigator stood looking out to sea. St Brendan, says his biography, written a few centuries after his death circa 577, set out to sea a few bays north of where the French fleet entered. He and his monkish crew journeyed west of Ireland and found various wondrous islands, which make a list as pretty as that of the French fleet: a barren island which turned out to be a resting whale who swam away when they built a fire on it but came back and became the island on which they celebrated Easter for the seven years of their travels; an island called the Paradise of Birds, where the birds, of course, spoke; an island of fellow Irish monks; an island of filthy giants who threw lumps of burning slag at Brendan and his crew; and, after many other islands, the Land of Promise. Some interpreters of the *Voyage of Saint Brendan* have asserted that the Land of Promise is North America, even to the extent of identifying the river flowing west the saint encountered, after walking inland for fifteen days, with the Ohio River. Others hold that, though the *Voyage* is merely part of an Irish literary genre of miraculous sea voyages, it does indicate Irish familiarity with the cross-Atlantic west in the early middle ages, and that other evidence – including the Norse sagas documenting pre-Columbian Viking encounters in the New World – suggests that the Irish preceded them.

*

I dumped my pack in the first hostel I came to, not red like Da Derga's, but pink: I had a room to myself with ten beds in it, each with a pink coverlet printed with huge cabbage roses. At dusk, which fell at half past nine or so in those late May days in that northern latitude, I went walking along the quay, and at its far end found about five trailers – or caravans as I later found most of the Irish called them – parked in an uneven row, the old-fashioned kind made to be towed by cars, though the cars parked nearby hardly looked adequate for the job. Some of these spindly pastel-enameled tin homes had generators sputtering nearby; one, gleaming and black like an idol, sat on a scrap of carpet whose floral pattern was almost blacked out by spewed oil. Another had a tiny chimney with a thin strand of smoke unwinding itself into the darkening sky. One front window was filled by a row of tall pink and white china vases with handles like swan's necks, vases that seemed to defy the unpredictable gravity of lives in motion, and most of the windows had lacy curtains. When I turned around at the end of the gravelly quay in what had almost become night and passed them again, two small boys in red sweaters were crying by the side of the road. A woman in a red van told the wailers, Get in the back and hide, and they scrambled in. As she grabbed a strap to brace herself against an abrupt three-point turn, her sleeve slid up to show an extensively tattooed forearm, and when the van passed me a man was sitting where she had been. They sped off, only turning their lights on as they reached town. That first glimpse of the Travellers was the only hurried motion I saw during my time in Ireland.

I had stayed up to listen to the blues in a tiny bar – very good blues, as it turned out, played by a handful of local youths who seemed to understand the melancholy of the songs as though it were their own. In the morning I woke up in my bed of roses and looked out to see a sky as cloudless as the heavens of home and all the buildings in the harbor waving gently upside-down in the bright calm water of the inlet. I looked again, and the gleaming image was gone; while I'd closed my eyes for what had seemed a moment, the tide had gone all the way out, leaving a bed of gravel, trash, and mussel shells. Nothing was open for breakfast that Sunday morning but two small hotels, and I was the only one who wanted to eat at nine in the one I settled on. They held me off with gallons of tea while the cook began the day, and then with piles of bread and butter. The food of Ireland could be called monotonous were not its essentials so inexhaustibly good, topaz-colored tea compounded with rich milk and demerara sugar and small loaves of bread, mealy and fragrant and crumbling, to say nothing of beer and whiskey. My brownshelled eggs came, and I cracked one and turned a pepper shaker over it. A fine brown dust floated down. It smelled so unlike pepper and so powerfully of hay and stables that I immediately saw the hayloft where I had learned to ride, the haybales in it, and the dusty path and the dusty oaks around it, the corrals and pasture beyond, a scene I had hardly visited and certainly not smelt in twenty years. I lost my appetite for the eggs, but the scenes the pepper had brought back hovered around me that day.

Bantry to Glengarriff was longer than the day before, but not as interesting, a long sinuous loop around Bantry Bay to its northern side, much of it with pleasant vacation houses to my right and the bay to my left. But for much of that day, I moved through the landscapes recovered in a dusting of pepper. From six to fourteen, as I've mentioned, I lived on a street full of suburban houses that was the last frontier of suburbia, and up the road called Seventh Street, on which our cul-de-sac was just a final spur, the country itself began. The worlds I lived in before that one were too vaguely remembered to leave more than a few epiphanies, the crumbling foundations of my memory, and the worlds after that came when my mind was largely formed by the world around Seventh Street.

My mother was from New York City, my father from Los Angeles; they had nothing to teach us about the countryside past the fading of the road, and perhaps the place was more vivid to me as one where things often came before words and where no adults arrived to interpret and regulate. In retrospect, my world then seems inside-out: home was an explosive place, and the hills Seventh Street took me to were a refuge in those days when children were allowed to go where they pleased. Children see with a peculiar intensity of vision; it is rare that a new sight or object can convey to an adult the hallucinatory power in which recollected experiences are bathed. What the very young see is literally incomparable – nothing like it has come before – and these encounters are the raw material, the imagery of their psyches. It often seems to me that all one's creation is done in that first decade and a half, when an internal landscape comes into being with the force and activity of primordial volcanos and plate tectonics; the rest of one's time on earth is spent retracing, mapping, deciphering, excavating. Everything else one will see is seen in comparison with this formative landscape.

I am bemused that such an Eden could wrap all around the chill of my childhood home. Childhood is often such a mixture of wonder and horror, a world the child is born naked into and out of which she makes a world around her – a more amenable one, with luck, and I have had luck. The outside world is no larger than the pores of one's senses, but the internal one is vast and shadowy, full of everything that has formed, been remembered, much that seems forgotten. Seventh Street, and books, are the first elements of the world I chose, my first refuges from the world into which I was born; Seventh Street is where I revert most often in my dreams, the road to anywhere and everywhere, the road all roads resemble.

The eastern side of Seventh Street sloped down, and a stud farm took up much of its length. Directly behind our back yard the old stud himself was pastured, a chestnut quarter horse of dignified and incurious bearing, even on those occasions when he dutifully heaved himself atop some mare. The western side of the street rose up in a small steep hill, on whose ridgeline was a blasted oak with an enormous dead limb resembling a stag leaping against the sky (when I returned recently it surprised me by looking that way still – twenty years is not a long time in the life of an oak). Up on the ridge itself were rocky outcroppings covered with lichens, the seats on which in my first years there a brother and I devoured stolen candy and in my last, we smoked pot. The local kids occasionally went cardboard-sliding – what coastal Californians have in place of sledding – down the hill when the grass was slick and dry. The old pieces of cardboard would moulder on the hillside along with abandoned lumber and fallen limbs; under them would develop grassless dark, damp patches in which centipedes, skinks, and alligator lizards lived.

These were neither so common nor so interesting to us as bluebellies, however, and a huge procession of bluebelly lizards passed through my childhood. I was always annoyed that the *Golden Guide to Reptiles and Amphibians*, the bestiary of our kingdom, called them by the plebeian name of Western Fence Lizards. They had a penchant for fences and other sunny spots, true, and their backs were a dustcolored mosaic of tiny jagged scales, their eyes like tiny ballbearings, but two bars of purest azure ran the length of their bellies on a creamy ground, a blue as pure and elevating as the sky. Snakes too were abundant in that landscape, the huge gopher and king snakes my fearless middle brother caught, and the tiny ringnecks I loved, slate-colored little things hardly thicker than a child's finger and perhaps six or seven inches long with a thin coral band just behind the head.

The North American continent begins small and intimate in the eastern seaboard and seems to gather scale as it moves west of the Mississippi, across the vastness of the prairies, over the high Rockies, the sublime expanse of the Great Basin, and the higher Sierra Nevada, but as it rolls close to its other edge, it begins to reduce its scale again, at least along the central California coast. With its intricacy of small, steep hills and rocky outcroppings, its dairylands and wildflowers, this my homeland reminds me of Ireland, at least during the two or three months of the green season. But its differences are profound too: the sky above was a high burning glass, not the sodden gray that keeps Ireland green, and the dry grass was crawling with reptiles. These reptiles lived in a world that was for most of the year dry, from the fresh gold dryness of late spring to the rain-leached iron gray of winter, and then for a few months at the end of the rainy season the whole thing turned a glorious green. A little farther up the road was the first house on Seventh Street, and one year the neglected grasses in its front yard grew so tall my younger brother and I trampled a maze in it. In places I could twist the grass overhead into arches, and then one day I almost stepped over the biggest bull snake I ever saw, almost black and infinitely smooth in the labyrinth of strawlike grass.

I never saw foxes or coyotes then, and I only heard rumors of mountain lions, but skunks and raccoons would wander down into the shrubbery of the subdivision. Sometimes early in the morning a deer or two would come trotting stately along the middle of the street as though on an inspection tour. There was a girl named Joy who lived near the top of Seventh Street alone with her father and a white horse in a field with a pomegranate tree, past the estate whose grapevines I also raided annually, along with sundry plum trees, prickly pears, and blackberry patches. It must have been the estate that planted the tall pines which flanked a stretch of the road and under whose low branches I built less a tree house than a field nest, of gathered grasses. I seemed always to be making little homes in the hills, finding hollowed trees in which to store treasures, rocky nooks to spend afternoons in, climbable trees. The best of all was at the top of the street, near where the pavement ended: a huge oldfashioned rosebush which had been running wild for decades until it was a mass the size of a large room; there was a low tunnel to its center, which was not a trunk but a cavernous hollow. Some nights I would lie in the still-warm grass of the hillside, my weight spread so evenly against the earth itself that gravity seemed hardly to hold me, and as I stared up at the stars the sky seemed a deep well I was hanging over and might fall into at any moment. The sensation of fearful vastness was my first introduction to the pleasures and terrors of the infinite.

Home, the site of all childhood's revelations and sufferings, changes irrevocably, so that we are all in some sense refugees from a lost world. But you can't ever leave home either; it takes root inside you and the very idea of self as an entity bounded by the borders of the skin is a fiction disguising the vast geographies contained under the skin that will never let you go. It is, if nothing else, the first ruler by which everything else will be measured, the place by which other places will be found hot or cold, bustling or serene, lush or stark. When I think back to my formation, it seems that landscape shaped me, made a home in the truer sense than the centerless house in the subdivision and an identity surer than the vague hints of familial and ethnic history that came my way. I am even literally made of the California landscape, of all the produce, water, wine I have been devouring since I was four.

What was passed along by my mother is of course a question whose answer approximates everything, from speech and language to her concern with social justice, which may itself be a legacy from her Irish republican forebears. She kept her emergency stash of money, in those days before electronic banking machines, in an old green copy of Liam Flaherty's novel *The Famine* and named her last child David, not to please her Jewish in-laws but after her brother and father who were named after the Irish nationalist and poet Thomas Davis, because her grandfather was an ardent nationalist (according to family legend, he was a Sinn Feiner who left Ireland under an assumed name, a wanted man). There are cups of tea and minor markings of St Patrick's Day and a certain kind of sentimentality I can pin to Ireland, anxieties about error, justice, sex, and the body I can claim were Catholic, but little more. Which is not to say that profound things were not transmitted so much as I cannot name them or their difference from the larger culture. My mother and her brother are confidently and wholly Irish-American, educated in Catholic schools, baptized, catechized, and confirmed. Their interests, their sentiment, their claims all lie that way, toward this poor island they have visited several times each, sometimes to track down umpteenth cousins and see homesteads of ancestors whose names they never taught us, who are no longer the children of that now foreign country.

The mad ex-San Franciscan Irishman in Djuna Barnes's novel *Nightwood*, Dr Matthew O'Connor, remarks that the Irish are as common as whale shit on the ocean floor, and in a country where nearly forty million people claim ties to the same island it is hard to say what distinguishing marks set the heirs of Irish immigration apart. Everyone in Ireland would demand my ancestry – or rather whether I was Irish, and if I said half, they would say, What's the other half, then. Inevitably, they'd want to know in which religion I'd been reared and when I said neither, they'd goggle slightly and say in tones of gratified horror, You mean you're *nothing*? Their surprise and my chagrin were signs of how far apart we were, I from the hybrid sprawl of California, they from the homogeneous Republic of Ireland. It was clear to both of us here that to be of Irish extraction was not at all like being an inhabitant of this supersaturated little island, that a trickle of that mythological substance blood had little to do with being reared in a soggy green country with too much history and not enough industry. But they would often assert that I looked Irish, confirming something to their own satisfaction, though Eastern Europeans are always happy to trace in the same lineaments signs of their region. The two inheritances seem rather to cancel each other out, making me as much double as nothing, a compound that could never be divided back to its constituent elements, a Jew to Christians, a Christian to Jews, a European to Native Americans, and an American to Europeans. To truly go back where we came from, I would have to be dismembered and divided among many lands, and my heart, I was sure, should be buried in the California hills.

There are fashions in remembering and forgetting: the melting pot assimilationist ideal of much of this century celebrated jettisoning the past to reach more quickly a utopian future like a shimmering white city; while the recent reaction against this restless forward lurch has stressed roots, blood, ethnicity, difference, remembering, a past as dark as dirt. The place of the tangled present, and of place itself, seems never to have surfaced in these schemes built on hope or on history. And the conversation of ethnicity rarely speaks to the hybrids and mutations that go on in a new soil, and it was this new soil that seemed as much a parent to me as the people in the house below the hills in which I grew up. In the endless succession of intellectual fashions, bioregionalism – a philosophy of coming to belong to one's locale by coming to know and respect its history and nature – seems to have fallen out of the conversation as multiculturalism made its entrance. So no chemistry proposing identity as a compound of ethnicity and geography has been made, nor any balance struck between roots and restlessness. Seventh Street I dream of still, one of only a few real places to take on any permanence in the territories I wander in my sleep. And Ireland for me was a place that looked like it, a road that extended from that first road.

Glengarriff was scattered along the main road, which dipped down toward the sea after shrugging off its few stores and pubs. In the Cafe de la Paix, I got tea and an egg sandwich and sat out front, watching the life of the town crawl by. Almost immediately a stout ruddyfaced man walked up to me, one of a long succession of garrulous middleaged men who would buttonhole me and who seemed incapable of talking to me for more than five minutes without touching on Irish history. These men – professional Irishmen a friend calls them – always intimated their detours through history were for my benefit, but they seemed to be for theirs: history was an itch they couldn't help scratching. In 1835, Alexis de Tocqueville met one on a stagecoach, who spoke of what had happened locally since Cromwell "with a terrifying exactitude," an exactitude that meant the wrongs were still felt and retribution was still due; and when the British Prime Minister Lloyd George met with Eamon de Valera in 1921 to discuss a peace treaty, de Valera annoyed him by harping on Cromwell, who was ancient history to the former, but not to the latter. Perhaps history is too fine a word for it: history is a continuum of innumerable facts;

mythology is the pattern of origins and reasons one picks out of them. The remembrances of professional Irishmen tend toward nationalist narratives in which their ancestors are victims and heroes, pure and simple.

This redfaced man was traveling with his sister and niece, he explained, but they stayed under their own umbrella. Where was I from, he inquired, and when I said California, asked if it was San Francisco, and then, settling himself in, asked whether I knew the Abbey Tavern there. We paid our mutual respects to the Abbey Tavern, and then he knew all he needed to know about me, and began to tell me all I needed to know about everything else. His conversation roved over a vast terrain, anchored every so often by a statistic. He told me that though it's only 300 miles from Mizen Head in the south to the far northeast of Northern Ireland, the country has more than 3000 miles of coastline; that in the Galway village he was born in, the policeman's wife had 18 children and the tailor's 24; that of the 8 in his own family, 5 are no longer in the Irish Republic, and he'd worked abroad himself for many years. He waved his hand at a cluster of adolescent girls bobbing by and said, They'll all have to emigrate to find work – England, the States . . . and Germany now. I don't know how my redfaced man worked in that he had 3000 books at home (one for every mile of Irish coastline, perhaps) or how many languages have died in this century – 1000 (I guessed correctly which one had been brought back to life: Hebrew).

And then suddenly he was talking with mild personal grievance about the Penal Laws, put into effect at the beginning of the eighteenth century. He enumerated the laws restricting land ownership, the sporting of certain haircuts and garments, the practice of religion, almost all forms of education, and said that the Irish Catholics had been like animals, no, even animals are allowed to be themselves. And with a brief celebration of Daniel O'Connell and Catholic Emancipation, which came in 1829, he returned to his companions, who were quietly sipping their drinks. I'm surprised, in retrospect, he didn't see fit to tell me about O'Sullivan Beare too, whose encounter with the English and the Irish weather (Ireland's history is partly a history of unlucky weather) was more brutal than the French fleet's in Bantry Bay. Defeated at Dunboy Castle along the northern side of Bantry Bay in January 1603, he fled from Glengarriff to the Curlew Mountains two hundred miles north with a thousand followers, including many women and children, on what became famous as his winter march across Ireland. Only thirty-five survived the weather and the enemies along the way, and their route was said to have been marked by the corpses of the rest of the party. O'Sullivan Beare himself was among the survivors, and he sought refuge in Spain. A few years later the Flight of the Earls brought further members of the old Celtic aristocracy into exile in Rome, and with that their reign was over.

I stayed that night in an old farmhouse partway up the mountain over which they fled, among rocks and woods from which rabbits shyly emerged to crop the grass at dusk. There was no one else there but a strapping American girl of about twenty on her first European journey and the proprietor, a sad, potbellied little man trying to make a living by taking tourists around in a boat by day and letting them stay by night in what had, in his father's day, been a working farm on the steep slope. The girl seemed like a Henry James character, one of those freshfaced incarnations of American naivety and force, and she spoke of all the marvels she'd seen in an odd metallic voice that seemed to have something to do with the long scar around the base of her throat.

7 *Wandering Rocks*

A pirate sails around the world in the name of his Protestant queen, looking for gold, either the gold of enemy ships or the gold of exotic half-imagined lands. He finds it, sometimes, and loses it, sometimes, and ventures as far as the western side of South America in his boat made of English or Irish oak. Peru has already been raided, and the Incas robbed of much of their glittering wealth. During an expedition of 1586, considered a failure because he missed the gold-laden Spanish fleet by twelve hours, the pirate probably – the record is murky – brings back something more modest, in fact the most modest thing imaginable: a lumpy brownskinned tuberous root that had formed the staple of highland Peruvian diets for centuries.

On another long voyage seven years earlier, he stops on the west coast of North America, at a place whose rolling hills and oaks remind him of his own island home, because he names it New Albion and claims it for his queen, who will always remain oceans and continents away from this conceptual claim. The land already has many names, among them the Spanish name *California,* after a mythical island of women warriors, though neither the pirate Sir Francis Drake nor the Spanish know it's not an island, and his name doesn't stick. His warrior queen is already busy subjugating peoples nearer at hand, the Catholic Irish and the eastern tribes of this continent he's landed on, and the strategies for both are much the same: plantations, settlers, massacres, and treaties made with men not in a position to make them and then enforced.

The potato perhaps brought back by Sir Francis Drake to northern Europe becomes a staple everywhere, spreading first through Europe and then even into Asia and becoming so much a part of local fare few remember whence it came. No one becomes so dependent on it so quickly or so completely as the Irish and, as the potato economy enables them to become more populous and more impoverished, they begin to cultivate a single strain called the lumper, large, unappealing, prolific, and easy to grow. A monoculture is a dangerous thing, but because of the instability of a wartorn place, a crop that grows under the ground, requires little cultivation, and can be stored as subterraneanly as it grows is irresistible. The single strain succumbs all over the island to a single blight, rotting overnight in the fields in 1845 and 1846 and on and on for a decade, and the Irish Catholics who have lived on lumpers and died of politics leave off their ferocious adherence to the local and become a race of emigrants, emigrating particularly to North America, to the English-speaking country that has shrugged off the mantle of English empire, to the United States.

Both of these colonies, first Ireland then America, become refuges for the Scots, who cluster most thickly in Ulster and the American South. In Ireland they never become Irish, and as the centuries pass they think of themselves more and more as British (though the Scots were Celts, and originally from Ireland), until Ireland's time of liberation draws near and they become the Unionists who succeed not in halting the revolution but in making it incomplete. A quarter of Ireland remains part of the British Empire, a last squalid little corner of what was once so far-reaching an imperium, and in this corner the liberation struggle deteriorates into a routine of revenges. But the Scots Irish in America are another story; many of them go to the southeast and become the most settled white people in the nation, the ones with traditions, a relationship to the land, a collective memory – with all those elements that constitute local culture. They make their whiskey out of corn and their houses out of wood, but they sing the same haunting ballads, ballads which are, as Bob Dylan once said, "about roses growing out of people's brains and lovers who are really geese and swans that turn into angels." Even in the ballads written in this new country, even late in this century, the romantic, violent, topographically specific, and fatalistic strains of Celtic legend gleam, and the dialectic of homesickness and restlessness fuels thousands of miles of song.

This country where they become free is the same one where Africans become slaves. In their homelands, they were West Africans and West Europeans whose identities were determined by culture, heritage, region, but in this mixed new country, skin itself has currency as meaning, and they become blacks and whites. The whites who were at the bottom of the social ladder in Europe now have someone lower than them, and a lot of them seem to like it that way; they live for centuries in highly structured suspicion and interconnection. The ballads and rhythms of their musics mix with least inhibition, and in the twentieth century new indigenous musics evolve, out of the red dirt, the strong African and maybe Native American beats and rhythms, the Celtic melancholy, into the hillbilly music cleaned up as country and western, and into blues and rhythm and blues. They all dovetail as rock and roll, a medium that spreads less like imperialism than like the potato and becomes a local crop all over the world, particularly the English-speaking world, a local crop that expresses the insurrection of the young against tradition and authority, of the margin against the center, and that sometimes becomes an institution itself too, like U2 in Ireland. The melancholy and the exuberance of slaves and outsiders have come, or come back, to Ireland, and by the time I wander across its expanse not only rock and blues but country and jazz bands are playing, along with traditional Irish bands, in all the small towns I reach, and it is possible to eat fish and chips and hear live blues in the poky town of Bantry on the west coast of Ireland.

Bantry, where in 1796 Wolfe Tone's rebellion failed. Had it succeeded, Timothy Murphy might not have jumped ship in northern California around 1828, not far from where Drake landed in 1579, and if the Famine hadn't happened perhaps my mother's four grandparents would have stayed home in Ireland too. To imagine Timothy Murphy as a personality bracketed by those two countries, it's important to remember what they were then: Ireland had ninety more years to go as part of England, and California had a decade and a half before it ceased to be Mexico and became the United States. The former was densely inhabited by its poor Gaelic-speaking indigenous population, lorded over by the immigrant English; the latter was in some ways similar, though its indigenous population is estimated never to have exceeded about a third of a million, a population broken up into about a hundred languages and cultures as befit the wildly varied terrain of California, desert, mountain, forest, grassland (California is five times as big as Ireland). The Irish, like the original Californians, had been too fragmented, despite their cultural homogeneity, to rally together against invasions on any grand scale, and so they became a conquered people.

For the Californios, like the Anglo-Irish, the 1830s were something of a Golden Age; they lived as rural gentry atop a huge labor force that was not exactly enslaved but far from free, they spent their time on dances, festivals, horsebreeding, and racing. These Mexicans and their indigenous servants were remarkable for their horsemanship; the saddles with their high cantles, horns, and long stirrups came from Spain, but the utter refinement of skill with which the vaqueros rode, roped and cut cattle had no European precedent. Cowboys didn't start out as the kind of white, English-speaking Protestants that populate most Western books and movies, but their Hispano-indigenous origins west of the West have long been forgotten. In 1846, Irish Catholics began pouring into the eastern US in flight from the Famine and Catholic Mexico began fighting a losing battle for its northern half. Some Irishmen joined the Mexican Army in what became the San Patricio Brigade.

After the war ended in 1848, the Mexicans joined the native Californians as an underclass with roots in a land that was no longer theirs, and then many other Americans who had already forgotten how California came into the Union began to denounce Spanish-speakers as outsiders. The more radical of those speakers still call California occupied Aztlan, after an indigenous name for what became Mexico, but some of them also mention that the same Spanish that is the language of the colonized in the US is the language of the colonizers in Mexico, a forked tongue. During the First World War, Germany pursued a policy of trying to create local trouble for its enemies; it was for that reason they encouraged Irish nationalism and the 1916 Easter Rising. They encouraged Mexico too. The United States entered the war on the side of England after the infamous Zimmermann Telegram, an intercepted diplomatic cable in which Germany proposed urging Mexico to try to regain the territory it had lost in 1846–48.

But back to Timothy Murphy, a Catholic Irishman, in the halcyon days of the Californios, after the atrocities of the missions initiated and before the atrocities of the Gold Rush carried out the decimation (but not the extinction) of the indigenous population. California was a nominally Catholic country before 1846, and most of the Americans who also settled there before the US seized it were Catholics or converted to Catholicism to marry local women. Murphy was a huge bear of a man, said to have weighed about three hundred pounds, and as Don Timoteo Murphy he became a great favorite among the local rancheros. The administrators of the region gave him land that wasn't really theirs to give, except perhaps by right of conquest: a vast estate covering much of what is now northern Marin County between San Rafael and Novato, the county on whose coast Sir Francis Drake landed and which he named New Albion and claimed for England, names and claims that didn't stick, the county I grew up in. There is little to see of that era, when local grizzlies were set against imported bulls for entertainment, a dozen families owned the whole expanse, Murphy raised wheat to sell to the Russian fort a few dozen miles up the coast, and the local Miwoks had not yet forgotten their language or lost all their land and their names for it.

Murphy never married, and he left large plots of land to the Catholic church. When I was growing up, the old Catholic cemetery in the next town south and the beautiful church and orphanage out in the country were still in operation, fruits of his generosity that far outlived him. The two of them, half-abandoned, with Italianate baroque architecture, overgrown weedy foliage, and the aura of romantic ruin, were the closest thing to the Europe I longed for then, and I cherished them. But there were no stories about them, or about Murphy, or about much of anything, and so I looked east for history. For all the passion I had for my locale I saw no way of staying there and uniting stories with landscapes, and an unstoried landscape is still somehow not yet a place, a home, a ground. I moved to Europe when I was seventeen, with money I saved up for years, beginning with my first job, in a used bookstore called New Albion Books on Sir Francis Drake Boulevard. Such yearnings sent me the first few times; later a passport and a story about a migratory beggar propelled me back. (Everyone's life is tangled up in story-lines that stretch far around the world, and if mine has any significance at all, it lies in ordinariness, not exceptionality.)

The sense in which Ireland was a colony is easy, concrete, political; there's a sense in which my real homeland is also a colony, a cultural one of Europe and the East. A colony is, most simply, a place and a people forcibly told that they are not central but marginal, that history is elsewhere, about someone else – perhaps a place where memory and history don't coincide. I have a permanent antipathy to the New England poet Robert Frost simply because I was assigned too many poems about spring, and snow, and maple leaves as a child, terms which at first seemed to me like nothing more than poetic convention, the same poetic convention that showered me with cuckoos, bluebells, duchesses, and other such unfamiliar things. Where I come from we don't have four seasons, and the imported maples are so confused they often turn red in January, when the plum trees are beginning to bloom. Even the cheap motels across the West I know so well inevitably have some degenerate descendant of an English Gainsborough or an eastern Hudson River School painting on their walls, images of a lush, domesticated rusticity that has nothing to do with the arid expanses outside, evidence that the locals have not yet learned how to imagine the place as home. Even the wars we were taught about had happened far to the east of us, and the war that united our vast region with the United States was glossed over as a minor detail. This is, of course, slim stuff in comparison to real colonialism. Ireland's own landscape was squeezed and scraped and made to yield profit, so that it never had much of the leisure and rural felicity of the dominant culture's ideal landscape. Likewise, until independence its schools too taught history as something that happened elsewhere, to other people, though its people were better equipped to counter the official version.

I want a Garden of Eden in which the serpent is a diamondback rattler and the fruit of the tree of knowledge is a prickly pear. I didn't find it then. The natural seemed utterly separate from the cultural in this landscape whose history no one could narrate and whose relics had all been ignored or effaced. Toward the end of my days in Novato, when my parents had split up, my father briefly dated an amateur archeologist, who told me that in the 1950s the biggest Indian burial mound in California had been bulldozed into flatness without being excavated, to make way for one of the shopping centers in the center of town. With that a place familiar to me all my life became mysterious, the site of an obliteration. Suburbia has ever since then seemed to me a voluntary limbo, a condition more like sedation than exile, for exiles know what's missing. But we were on its outermost edges, and other worlds came seeping in.

Go up Seventh Street – the last time for you, one of countless times for me. Go past the horse pasture to your right and the hill to your left, past the house whose tall grass made a maze, past the plum trees and the pines, past the huge prickly pear cactus I dreamed about once in a dream of flying away from a man who was two men I loved, past the rosebush cavern, past the old bottle dump we excavated, the buckeyes, and the shed, over the crest of the hill and down the gravel road past the pasture where I learned to ride, through the dust past the end of the road and down to the marshy far end of the pasture where the sticky weed that smells like tar and licorice grows, through the loose strands of barbed wire, across a wider cow pasture to the foot of the biggest hill between my town and the next. The hill is named Burdell, after a rich and insignificant dentist who lived at its foot a century ago, but the place where he lived is called Rancho Olompali. This is where I'm headed, to the war at the end of the street.

Only the hot days during summer vacation seemed long enough to reach Mount Burdell, the highest point between Novato and Petaluma, a big hill around whose head oaks clustered among the grass that would always be bright, dry gold by that time. There was a spring halfway up, welling out of a little cement basin, and a quarry with all the charms dangerous plummets have for children, and scattered deer waking from their daytime slumbers to burst away in weightless leaps. Running along the top of the mountain was a long wall of stones gathered from the slopes around (built in the nineteenth century by Chinese labor, as I read later, but most of those laborers had gone back to China or died unmarried and forgotten; the wall is their only monument, as the shopping center is a secret memorial to the erased cemetery). The wall marked the boundary of Rancho Olompali, whose buildings at the foot of the hill faced the northern end of the San Francisco Bay. I saw it often and toured it once when I was tagging along with my brothers' cub scout troop. Otherwise no one mentioned the rancho in our history lessons or mentioned that any of California's history was local and tangible. When we peered at the mud bricks of the original adobe through a glass window set into the walls of the Burdell mansion that had been built around it, age itself rather than specific characters and events were supposed to make this dirt significant.

The site had originally been a Miwok village. In 1843, a Miwok man, Camilo Ysidro, who had successfully assimilated into the Spanish lifestyle of the rancheros (and whose daughter later married a Harvard man, a gringo), received the Olompali region as a land grant, much as Murphy received his huge spread slightly to the south. There wasn't much said about the Miwok either; in the days of my childhood the textbooks said that all California's Indians had been primitive Diggers with little culture, material or intellectual, and the same colonial textbooks implied they had conveniently vanished in some manner as vague and uninteresting as the Diggers themselves. I was an adult when I found out Olompali was the scene of the only fatal battle in the Bear Flag Revolt of 1846.

By 1846, it had become clear that the US was going to have California one way or another. The annexation of Texas was a prelude. There had been a few abortive raisings of the US flag in California in 1844 and in March 1846; and even the Mexican military administrator of the north, General Mariano Vallejo, was prepared to hand it over peacefully. The American consul in Mexican California wrote on 23 April 1846, "The pear is near ripe for falling." How ripe the consul could not have known: the first battle against Mexico began on the Rio Grande in Texas two days later. On 14 June, a group of Yankees, many of them in California illegally, began their own war by taking Vallejo prisoner at his Sonoma house, ignorant both of the war already being waged and of Vallejo's acquiescent position. Afterwards, Murphy came to help take care of Vallejo's family, and when he admitted he was a Mexican citizen he was taken prisoner too. He was released when he allowed his captors to add his name to the list of the conquerors. The Bear Flag Revolt, named after the flag they raised soon after their initial raid, was something of a farce, a handful of eclectic, grimy, and often drunk Americans chasing after the widely scattered Mexican citizens of the region (who thought the bear on the original flag looked like a pig).

On 24 June, the insurgents attacked the Mexicans at Rancho Olompali while the latter were eating breakfast—perhaps California's first drive-by shooting. It was the only fatal battle in the farcical rebellion, in which one Mexican died and a vast slice of the continent changed hands, or flags, or deeds. Long afterward, the marks of bullets were said to be visible on the old oaks and bays clustered around the house, though I have not found any. When I was taken there as a child, Olompali had passed into the hands of a real-estate consortium and then become a hippie commune called The Family. The late Jerry Garcia of the Grateful Dead once said that he had his last LSD trip there: "He developed three-hundred-and-sixty-degree vision, died a few thousand times, and saw the word 'All' float into the sky before he turned into a field of wheat and heard 'Bringing in the Sheaves' as a coda," explains a biographer. In 1969, The Family accidentally burned the mansion down, but the adobe walls survived. Its trajectory seems to embody California history perfectly: indigenous village, mestizo rancho, Mexican-American battlefield,

Anglo-American estate decorated with exotic plantings, abandoned psychedelic countercultural community. Since then, the place has passed into public hands as Rancho Olompali State Park, though its significance as a war zone still goes virtually unmentioned, as does the war itself.

The California Grizzly pictured on the flag these robbers or conquerors raised, the flag that is still the state flag, was hunted into extinction by the early part of this century. Soon after the Bears declared California a republic, the war with Mexico expanded to the Pacific front, where it dragged on until the Treaty of Guadalupe Hidalgo in 1848. The transfer still holds: California is part of the United States. No one seems to remember that all this happened; history was something in books about someplace else; and I was waiting to grow up to visit countries hallowed by history, the history I knew best, western European history, particularly English and French history. Now I have unearthed the history that seems more mine than that of a faraway continent, the long largely unrecorded oral and brief bloody written history of the American West. And I have questions, questions that rise like fumes from the grounds of these places, the very ground I stand on, which is a graveyard as much as a garden.

If Ireland properly belonged to the Irish after seven and a half centuries of English occupation, who does California belong to? If some of the English became Irish and thereby belonged to Ireland even if it did not belong to them, what is the process of becoming native in a place where *native* is such an ambiguous term? What about all the rest, not the Scots Irish whose interim presence in northern Ireland became a real relationship to the American South, but the Catholic Irish who came later, after the failure of the potato, and who so often decried English but not American imperialism? If being an Irishman meant being a colonized subject, did Timothy Murphy cease to be Irish when he became a colonist? Did he become Mexican, then American, as potatoes and rock and roll became Irish? Does being American require a more profound degree of adaptation, a passionate knowledge of place that is abysmally lacking in most Americans? Such histories as these suggest that identity is not a social but a geographical science, and they suggest that the opposite of remembering may not be forgetting but creating, out of the mixed and hybrid materials that come with relocation.

There are still distinct places called Ireland, and Peru, and California, and England, but only two languages prevail there now, Spanish and English, and potatoes and blues and bloodlines have so mixed up the contents of these countries that ethnicity without geography means something else or means nostalgia for a time when things were simpler and separate, for the time before even rhythms and spuds went rolling round the world and words like *home* and *native* were easier to say.

8 *Articles of Faith*

A pair of feet, moving along a road. Feet imply legs, and on and on, to a head, with histories inside, and the feet are mine. A road implies landscape, and around this road, steep and heading north, are thickets and steep slopes screening the rest of the earth. So far, the world consists of nothing but me and the road, the narrow road that eternally snakes around another bend, over another crest, out of sight; the me that consists of a bevy of sensations and thoughts and no sight but the swinging of hands and feet into the periphery of the view – of the road. Everything else in the world is a leap of faith, and I am only walking. To say that this walker has a name, a past, another life, has ancestors, that half those ancestors came from an island called Ireland, that this place with the island-name is what spreads far to the north and east of this road – those things are articles of faith that I sometimes forget, sometimes believe, but I'm not sure I know, though I know the names. Everything I'm sure I own is on my back, or in my pockets, for the world I've come from may have been destroyed, and I wouldn't know it, out here alone. I'm traveling as light as I can. One can leave the words *home* and *native*, words that are about staying still and remembering, out of one's load and try being nothing but a traveler in a landscape bisected by a road – and a road is not itself a place but the way between places and the ribbon that ties them together.

Being in motion wakes the body up. In repose one is nothing but a surface of potential sensation, only the surface, the skin, is awake. Exertion and pain make the rest tangible – otherwise bones and muscles and organs would be little but articles of faith beneath the visible and sensible surface of skin, and so one's own interior anatomy may be among the things explored in the course of a journey's exertions. At the same time, journeying reduces the self to the boundary of skin: everything else is foreign, unknown, belongs to unfamiliar others. In the course of many travels, I was beginning to learn how much the self, the soul, extends into its world, the world one calls home. Though always described as an autonomous, individual, internal thing, it inhabits the world that welcomes, shapes, and accommodates it. At home I was a creature insulated within a home that fit me snug as a shell, and beyond that a circle of friends who permitted and provoked me to think and speak of certain things, to draw forth certain possibilities that might not exist otherwise, and beyond that a society and a landscape so familiar they formed almost a second skin for a self that has only moved thirty miles from the road of my dreams and the house of my childhood. It's a pre-Copernican world of nesting crystal spheres, and among those spheres, ancestry and nationality were among the most distant constellations, faintest in the daytime of the familiar.

The world called home, often a world one has carefully constructed, continually summons a particular version of oneself into existence, the version made up of what one speaks of, does, sees, eats. In traveling this insulating construction is pared back to something so basic that such modifications and elaborations have not yet been called forth, or the unfamiliar world calls altogether unfamiliar selves into being. I remarked sometime before I arrived here that the external world is no larger than the pores of one's senses; the internal one so much larger, full of everything that has happened and been imagined. The opposite is equally true: the self sends out tendrils into its world and becomes far larger than the body, which is only its core. To know something is to have it enter the little unseen kingdom under one's skin, to love it is to have it become one of the enlargements of the boundaries of self.

If skin is a boundary, it is an open border across which a vast import and export business of sensation, secretion, passion, and consumption passes. Skin is a literal border too: the body is two-thirds liquid, a liquid that longs to spill into the larger world and reunite with its sources, and only skin keeps the body discrete, a self-contained country rather than part of the fluid substance of the world. Solitary travel radically reduces the self, pares it back to the proportions of the body, seals it inside the country of the skin. It's good to know what is portable, independent, what survives translation to an unknown country; and it's good to know that one could start over again, build up from scratch acquaintance with the languages, weathers, customs, roads, and friends that extend one into the world. Pared back to essentials on the road, one is also severed from the present and thereby free to wander through the past and taste other destinies. The rest lies dormant, unrecognized, an interior life that may not be called forth again by the outside world. The world one builds for oneself, whether prison or palace, is a world one can partially exit – not in the mind, but in the body – by traveling, by going to a place where nothing calls it into existence, where one is stripped back to the boundary of skin. Or so I thought.

So I was walking up a road, no one who knew who I was knew where I was, and I didn't know what lay ahead, though I did know that I was on the road that would take me across the border between Cork and Kerry. Or so the map said, and a road-sign said the next town was twenty-five kilometers away. Up high, there were no more trees, and I could see further – down to Bantry Bay, the islands around Glengarriff, and, far below me like a map itself, the checkerboard of scrub-gorse, stone, cleared fields, real forest, pine plantation, and houses of a model rural landscape. Sheep wandered freely across the road, climbing the tumbling stone fences at my approach and bleating mechanically. When most frightened, the lambs ducked under their mothers to take refuge in nursing as though they could return to the tiny secure world of the womb, and their tails jerked back and forth like furious metronomes in apparent synchronicity with their suckling. Twin lambs dived upon a ewe with her back to me, one on each side, and she looked like a grounded airplane as their tails moved like fuzzy propellers. I walked past innumerable tufts of their glossy white wool, smooth as hair and more wavy than wooly, caught on sticks or laid on the ground, looking like locks yanked from the heads of old women.

I walked, one foot then the other, connected and disconnected to the ground beneath my feet, toward the top, moving at the pace of my body, which seems idiotic to point out, except that it's a rare speed nowadays. Paddy remarked that before the Famine, the average Irish peasant probably never went more than ten miles from home, the distance of a day's walk and return. It suggests either a terrible monotony or an intimacy of astounding depth, an existence in which the world and the self were so interfused they would hardly seem separate phenomena, and self would be indistinguishable from its circumstances (which might help explain the extreme trauma of emigration for those who had not yet, as beggars and laborers, become migratory). Without machines (and horses, but who travels by horse now?) most human beings would be limited to about fifteen or twenty miles a day in progress across the surface of the earth, perhaps twice that for practiced ramblers and runners. Human scale is a term most often used in describing objects rather than motion, but the mind still doesn't absorb a sense of place at any speed much faster than the human body moves. Faster than that, at the speed of machines, the world becomes a stage set and only the largest changes register. Only residency entitles one to claim to know a place, but slow and leisurely progressions give one an inkling.

The motion of walking, this rhythmic rocking, seems to set the mind loose to wander on its own paths. The weight of baggage on my back, the weight of history in my head: the picture of a world bereft of everything but a walker on a road is nothing more than that, a picture, the picture I set out with, the fantasy of travel as an opportunity to clear my head rather than rearrange it. I went walking through concentric rings, the largest one called Europe, then Ireland, then the west, then the road to Kenmare, each with its own associations, towards literal rings, to stone circles. There are more than a hundred such stone circles in west Cork and southern Kerry.

The rocky ground became a ground of rocks: great slabs tilted and breaking through the earth, rough, a darker granite than the granite I knew. It was a big, wild landscape, and for the first time I felt satisfied rather than merely charmed with what I saw. At the crest the rocks rose up almost like mesas, with sheer vertical faces perhaps twenty feet high, on which lots of lovers, travelers, and one "Kentuckey chicken" had painted their names. They had no inkling of spraycan style, these brushy souvenirs of passage, and they looked instead like giant epitaphs. At the very top was a tunnel through the rock, full of drips and puddles gleaming in the stony darkness, and I walked through it. Out past the brief dark I arrived in another world, in Kerry, where a totally unexpected and even more magnificent view dropped out from under my feet on the heights, an inland view on the east side of the crest, high above farmlands nestled between this and further ridges in every direction. The sheep changed too; their haunches were no longer inscribed with red letters but with marine-blue *A*s; and I imagined that when enough flocks come together they must look like alphabet soup or a printer's type box overturned.

An hour or so later, down at the bottom of the slope, across a little river, I took the secondary road, because it passed, said another one of my red maps, a circle of standing stones I hoped to find. This road was truly remote, two parallel tracks of crumbling asphalt with a grassy middle strip on which I walked, hoping that it would please my feet. On this third day of walking, the shortcomings of my boots or the slight extra weight of the pack – twenty pounds of water, clothes, and papers – had begun to grind my feet down. Poor feet, intricate architecture of half the bones in the human body: I could feel every bone in them with every step, clearly enough to draw a map of them, an x-ray of pain. They felt like twin larks, these fragile metatarsal fans of bone being flattened and splintered more with every step, and it came to seem peculiar that the solid architecture of the human skeleton with its sturdy pillars for legs should have instead of a solid foundation the fine bones of birds at bottom. The last ten miles would be excruciating, but in summer's long light, there was no hurry. I walked, and noticed every step.

An old man looked up and said incuriously, It's a grand day, as he dragged some branches out of a pasture. He was the only person I saw, his the only inhabited house on the way up this slope. Water sang everywhere on it. Little Stonehenge-style post and lintel structures frame the trickles coming out of the hillside itself, trickles that shone as they emerged from the darkness. Everywhere in Ireland the boundary between architecture and landscape blurs, and sometimes it seems that the walls and lintels and ruins are wholly natural, sometimes that the earth itself is consciously evolving more elaborate structures of stone. The doorways into the hillside made it look inhabited, a subterranean home. The second pass was even more magnificent, abandoned even by sheep, flanked by waterfalls, wild again, and then it too descended into the rustic: another farm valley on the other side, with roads, the roar of a tractor, trees, and buildings. Down low the rough scrub gave way to hedgerow flowers: tiny warm purple orchids clustered on stalks, violets, vivid blue things like forgetmenots, and every once in a while bluebells, primroses, and foxgloves.

I hid my pack and turned off on what my map seemed to say was the cul-de-sac leading to the stone circle, past some houses and some barking, but then the road dead-ended at a farm tucked down in a hollow. Yips came from the farmhouse and more barks from a weathered shack off on the other side of the end of the road. An old woman came out, wearing a print apron over her lacetrimmed sheer nylon blouse and tiny red basketball shoes on her feet, and I told her what I was looking for: one of those small stone circles that are so scattered and unattended across the land of Cork and Kerry. She said she didn't think she'd heard of any stone circles in the area, but she was the kind of person people tell things to, but she forgets. Her husband might know, but he was off tending sheep, or the people up the road might. Where was I from then, she asked, and then she remarked on how nice and comprehensible my English was for an American and added that it was probably going to rain anyway – she had heard a thunderclap. So I thanked her and gave up and walked back to the car she thought I had. For me the country around her house was a picture labeled *Europe: case in point, rural Ireland, from the neolithic onward*; for her it might not be a picture at all, and the label would be something like *Where we live and labor: Home*, a landscape made not out of ancient artifacts but of contemporary sites of work and resources, where water comes from, where the sheep go, where she resided.

Crunch, crunch, crunch on the larks, downward to the slow-rolling Kenmare River, which disappeared again when I dropped among trees, and with a few thousand more steps, I arrived at the bridge across the broad river estuary in the town of Kenmare. Kenmare used to be called Neiden, which means nest in Irish, and it is a nestlike place, small and sheltered. It was still called Neiden, or Nedeen according to Arthur Young, when he passed through in September of 1776; Kenmare was a name applied by Lord Shelburne around that time in honor of his friend of that name, who owned much of the land to the north. Shelburne owned, says Young, "above 150,000 Irish acres in Kerry."

Young was a man with aristocratic ties and a consuming interest in agriculture; his two-volume chronicle of his lengthy journey around Ireland deals largely with its current and potential farm industries. The accounts would make fascinating reading if only for explaining what all those aristocratic men in the novels of the period would have been up to when they weren't playing whist or flirting in the drawing room. The conventions of fiction show us these men in their social rather than their economic roles, and so agriculture and the concomitant relationships between landowners and land-workers is rarely mentioned. But Young's is not a novel, and his treatise tells us they were captains of agriculture overseeing experiments in manuring with burnt limestone, seaweed, and dung, clearing fields of stone, working bullocks by the horns, draining bogs, and planting them to clover – if they were diligent and in residence on their estates. Irish landlords were famously absent, dissipated, and brutally exploitative, but a surprising number of them were at home to host Young and to show him around the farms on their estates. They did not always succeed in convincing him that the peasantry was lazy rather than oppressed – Young is also notable for his evenhandedness in deciding why the peasants were hungry, ragged, and abysmally housed. Lord Shelburne himself must have been off in a drawing room somewhere when Young came by, but his representatives seem to have been carrying out a fairly enlightened regime. Kenmare is still idyllic looking, and still shaped by the model cottages Shelburne had made.

The whole town appeared to be made up of nothing but pubs for the locals, restaurants and shops for the tourists, and picturesquely cramped homes. After a late afternoon repast – the usual tea and bread, unusually welcome as the first meal of the day – I followed the model cottages down the main street to a triangular town square. Two middleaged women there passed each other, and one said, Evening. Bit of rain, not a question or an opening or a revelation, only a fact. And when I went up a back street to the stone circle on the edge of town and into the alley that angled off it, I ran into a stout woman taking admission for it. Her nose and cheeks were a cluster of red bumps and she said, It's raining, and then that it rained forever, ceaselessly in the winter and spring, a complaint I heard often. Homing in on defenseless me, rocking on my mangled feet, she collected my fifty pence and imparted detailed information from her television about the weather in Scotland, in Waterford on the opposite coast, about everyplace on the island, and the showers expected for the morning. (Kerry is the wettest part of Ireland. When Noah's ark went by Ireland, Paddy told me, there was a Kerryman sitting on top of Mount Brandon, the waters all about him. He hailed the ark and asked Noah for a lift, but Noah turned him down, explaining he had his orders. Never mind, said the Kerryman, 'Tis only a bit of a sprinkle.)

Weather is one of the two perfect subjects for strangers, being perpetual and apolitical, and as I don't follow sports, I was confined to discussing the weather throughout my journey. Few neglected to tell me that it had rained for a hundred days straight before I arrived or how they felt about that.

Fortunately if it isn't raining it has probably just ceased or is about to begin raining in Ireland, and the variations on this theme nurture endless smalltalk. Weather is itself a perfect article of faith, being both utterly inevitable and utterly unpredictable. There's almost a ritual quality about weather talk; it may be that paying homage to its existence is how strangers acknowledge their common existence and vulnerability under the same unfathomable sky.

This one was dropping a light drizzle on everything as I proceeded down the alley, past hanging laundry and a drowsy liver-colored spaniel, to where it forked off into a construction yard on the right and a bank with a sheep-proof gate on the left. Up the bank and through the gate was a perfectly mown and weeded green lawn with fifteen rough boulders arranged around a central boulder. They were of two colors, gray and brownish, and all shapes, from slablike to rounded, and none of them was as tall as I was. The whole circle was small, perhaps twenty-five feet across. My compass made it clear the stones were neatly aligned according to the cardinal directions, and they also seemed to echo the mounds of the surrounding hills – but that was speculation. This and the dozens of other stone circles in the region, surmises the great scholar of such circles, Aubrey Burl, were built about thirty-five hundred years ago by a wave of invaders or emigrants who came from the coasts to the heavily wooded hillsides to settle – came perhaps from Scotland where similar but older and larger stone circles survive, but that's speculation too. "The builders penetrated inland up the river valleys until they came to soils light enough to cultivate with the equipment they possessed, probably no more than the spade," he writes, suggesting that each ring corresponded to a community inhabiting about one square mile of land, and that no more than thirty people would use such a ring at any given time – hence their diminutive scale, by the standards of English and Scottish rings. (When Young came through three millennia later, the plough had still not come into general use.) "These largely forgotten stone circles deep in the Boggeraghs or along the storm-drenched rocks of the western bays once were the most needed centres of people's lives, places that combined the mundane earth with the imagery of the night-sky . . . where people trod or danced along the portal stones . . . looking toward the darkening, lowering horizon and the setting of the sun or moon. Now only the stones and the secrets survive."

Stonehenge is impressive because its builders, possessing minimal technology, defied both stone and gravity to create such regularity of form and verticality of structure, with the massive lintels pressed up to the sky. But it also seems to imply, as does all more elaborate architecture, that the world lacks order and needs order imposed, that defiance is necessary, that an inhabitable world must be made rather than found. Such stone circles as the one in Kenmare are more modest, imposing only a minimal amount of order on the stones and the surrounding landscape. A kind of nonchalance permeates the site, as though such a circle of unrefined unmatched stones were good enough for the now forgotten purposes and its makers were uninterested in being any better. There's an ease in the sense of labor and of landscape such a casual grouping gives, and perhaps in a culture that could chart the celestial bodies with unimproved lumps from the earth like these. They testify too to an immobility which may have been the fixedness of agricultural society – and nothing could be more fixed in place than a boulder, save a boulder aligned with the heavens. A millennium later the Celts invaded and the precise functions of the stone circles were forgotten; for all the archeological discoveries of the last decades, the uses of such circles remain speculative. Like maps, like names, like conversation about the weather, they seem to have been, in the broadest sense, diagrams with which to navigate the recurrences and changes of the years, the tangibles which alluded to and kept track of the articles of faith.

9 *A Pound of Feathers*

Perhaps people travel for pleasure because the visual is much more memorable than the tangible, the seen than the felt. At the time, traveling may be nothing more than a series of discomforts in magnificent settings: running for the train to paradise in a heat wave, carrying an ever heavier pack in alpine splendor, seeing sublime ruins with stomach trouble. Yet it is the field of images and not the body of sensations that lingers. My mother once remarked that if women remembered what childbirth felt like, no one would have more than one child. And so I, third child of a third child, owe my existence to forgetting and my taste for travels to the dominance of the eye. I got up in Kenmare to find that a knee, a muscle, both feet were inexplicably wrecked as no previous walk had ever wrecked them, and they told me of it with every step. The marvelous walk over the mountains to Killarney which I had anticipated for three days turned into a bus trip. Small waterfalls tumbling over rough rocks and into sinuous streambeds flashed by, shadowed by oaks whose branches spread at the same abrupt and undulating angles, invaded by rhododendrons, all among a complex of passes that shifted the view around in unexpected, abrupt ways. The whole landscape had in its very lines a kind of intricate, tangled wildness I saw nowhere else in Ireland, and I had no chance to sort it out.

The expanse of countryside around Killarney has been considered, for two centuries, to be the most wild, the most romantic, and the most beautiful in Ireland. Some of this status is due to the place's inherent wonders, some of it to the fact that much of the rest of the island was deforested and domesticated long ago. What survived as forested parklands if not as wildernesses were gentlemen's estates. Much of the region – 91,000 acres – around Killarney had been granted to Sir Valentine Brown in 1620, said a plaque in the park; when Arthur Young came through a century and a half later, Lord Kenmare owned much of it, and a Mr Herbert seemed to hold the rest. All the elements of romantic landscape are here: steep slopes, long views, narrow passes, rough rocks, water as both placid lakes and turbulent falls, magnificent trees and groves, and the crowning romantic element, gothic ruins – all compressed into the intimate scale of Ireland. Young, well educated in the protoromanticism of his time, was distracted from his agricultural interests long enough to rhapsodize for many pages: "There is something magnificently wild in this stupendous scenery, formed to impress the mind with a certain species of terror. . . . From one of these heights, I looked forward to the lake of Killarney at a considerable distance, and backward to the river Kenmare; came in view of a small part of the upper lake, spotted with several islands, and surrounded by the most tremendous mountains that can be imagined, of an aspect savage and dreadful." Muckross Abbey, in Young's time, had already fallen into the kind of ruin so beloved of the eighteenth century, and the bones of monks were still scattered among the rocks. Kenmare and Herbert devoted themselves, in the fashion of the time, to improving the landscape with bridges and paths, making it more of a garden but not much less of a wilderness – by European standards, anyway.

The poet Shelley came to admire the scenery a few decades later, by which time the place had already become an established beauty spot, like the Lake District in England and the Swiss lakes. Tourism as we know it was an outgrowth of the eighteenth-century English enthusiasm for scenery, a taste that now seems wholly natural but was in fact a creation of the connoisseurs of paintings and gardens of that time. Their taste in gardens became more and more naturalistic, until, as Horace Walpole remarked of the first great landscape garden designer, William Kent, "he leapt the wall and saw all nature was a garden." The making and owning of vast gardens was an aristocratic privilege, but the admiration of existing landscapes could be a far more general pleasure, and as it became one, scenic tourism came into being. Scenic tourism – traveling for the eye – which is now such a vast industry and popular activity, and which was for all intents and purposes the foundation of the environmental movement's first stirrings in the nineteenth century, is often imagined by its participants as an innate desire as well as a universal hallmark of civilization. But it had its birth among such heights and views as Killarney as a very specific, cultivated, and class-based taste – an English, and to a lesser extent continental, import.

So it was perhaps peculiar that Killarney, when I arrived, was full of a more medieval kind of traveler: pilgrims seeking wisdom rather than scenery, or pilgrims of a sort. The International Transpersonal Psychology Association was holding its annual conference in Killarney, this one titled "Toward Earth Community," with, among all the presentations on psychology and spirituality, some environmental, Native American, and Irish speakers. The California-based ITA is an interdisciplinary organization with a fickle enthusiasm for "new paradigms" and sweeping ideas. Its members were mostly well-off white Americans who were entering their own middle ages, the men's hair thinning and the women's bodies thickening, clad in the purples and talismanic jewelry of New Age subculture. A scattering of European and Irish people were there too, along with a few young people I knew as friends of friends, and the Native American speakers. I hung out for three days while my nether limbs recovered and got caught up in the usual spirit of conferences – a gluttonous delusion that all the answers are being offered, if only one can sit and eat up enough information from morning till night.

I skipped out periodically, but came by often enough to hear some interesting things. There was a native Hawaiian woman, Mililani Trask, who spoke of how tourism was destroying her island's ecology, and a Hopi woman, Marilyn Harris, who showed slides of her part of the world until I felt so at home, or back at home, among the red rock and blazing skies of the southwest that Killarney's green and gray came as a shock to me when it was over. Two women, Winona LaDuke and Helena Norberg Hodge, spoke brilliantly of the effects of capitalism and globalization on traditional cultures. The speakers often represented third- and fourth-world cultures whose spiritual traditions are still intact and whose political situations are dire. They spoke of the need for activism and pragmatic change. The audience seemed drawn from almost the opposite situation, one of great material privilege and a sense of spiritual poverty, and they wanted to hear about spirituality and tradition as solutions, rather than the political forces that threaten them and the activism that might protect them.

Perhaps I shouldn't say too much about the conference or its audience. It may be that the flocks of New Age followers annoy me because in some ways I resemble them. They are most often responding to the same ambiguous situation of coming from an amnesiac, hybridized, commercialized, and confused culture. The diagnosis is often not bad at all. But the solutions are where they lose me: they become spiritual tourists, and I mean tourist in its pejorative sense here. They often want to pop into another culture or era and pick up meanings and identities like snapshots and souvenirs. The very way they look for alternatives embodies some of the most pernicious aspects of the culture they come from: the desire for quick gratification, a kind of globalizing control over other cultures, the segregation of politics from spirituality. Their pursuit of awareness often incorporates considerable obliviousness.

Part of it is a pervasive insouciance about the concrete politics going on around them and about the politics of their own conduct. Spirituality as a depoliticizing worldview seems to isolate those who see it that way, and I often wonder if some of what they're looking for in spirituality might be found in the communities and purposefulness of political engagement. Too, much of the spirituality they pursue and practice isn't theirs: "Spiritual traditions cannot be mixed and matched in ritual potluck," said one speaker, a university professor named Bron Taylor, "without degrading all of them." This appropriation has put the New Age movement at odds with some of the cultures they claim to admire. A year before this conference, the Lakota Nations of the northern plains had issued a formal declaration of war on the New Age movement and anyone else expropriating, hybridizing, and abusing their spiritual traditions.

I remember one talk in Killarney in which a man who was supposed to be addressing forest politics told us all to close our eyes as he led us through a visualization of a forest, and I wondered why we were all sitting there with drawn curtains when one of the last great forests of Ireland beckoned outside. While I was supposed to be envisioning a forest, I scribbled notes about how somewhere in the last few decades the mainstream of philosophical inquiry had been split into two incomplete and unsatisfactory channels best described as the New Age and the academic. The New Age movement tries to find meaning and belonging by blurring distinctions and differences, but to say, for example, that Hopi culture is just like Tibetan Buddhist culture renders both meaningless. It seeks some common ground that will reconcile all differences and establish a single absolute truth, and toward this end it constantly asserts the interconnection and affinity of various cultures at the expense of significant differences. It may be, among other things, a strategy that allows them to insert themselves in the picture: if Hopis and Tibetans are really so alike, then the ground is common enough to admit a few hybridized Euro-Americans too. The absolute unity of all things, which has meaning as a mystical vision arrived at after long travel, doesn't make much of a starting point, and proposing it seems to be a way of taking a short cut.

Meanwhile many of the modes of thought pursued in recent academic practice seem most often to triumph in making absolute distinctions. Their worldview is razor-sharp, but so are all the objects in it, too sharp to touch. The only pleasure remaining in such systems is that of yet more rigorously defining or dissecting an object and thereby demonstrating one's mastery over it, a mastery that depends on one's distinctness from it. The conversation about ethnicity has veered between these two poles: a sort of postmodern multiculturalism has pointed out that the universal brotherhood of man has too often been used to suggest that the rest of the peoples of the world should or do resemble modern European man. An alternate model of radical difference has emerged, along with the idea that one should not speak for the "other," for those culturally dissimilar from oneself (which in its ultimate form means one can talk about nothing but oneself), or appropriate from other cultures. Such a model is far from the New Age ideas that ancient secrets link so many cultures and they are all available for the spiritual pilgrim; in some ways the New Age isn't new at all, but a strange, half-humbled, hungry vestige of the old universal brotherhood business of a dominant culture. Somewhere inbetween these schismatic traditions it seems that it should be possible to think with feeling, to feel thoughtfully, to find a middle ground between the fuzzy and the icy. But the present moment belongs to these twin opposites wandering through the same dark wood. One party of wanderers argues that there are no distinct trees in the forest, while the other argues that no overarching forest arises from the accretion of trees.

There's a middle ground, a path through the forest, a sense in which language and images neither lie nor tell the truth, but provide pictures whose correspondence to what is depicted is always imperfect, but always capable of further versions and revisions – one can approach, if not arrive. Truth, says Nietzsche, is a metaphor we have forgotten is a metaphor, "a mobile army of metaphors . . . a sum of human relations which became poetically and rhetorically intensified, metamorphosed, adorned, and after long usage seem fixed, canonic and binding . . . worn-out metaphors which have become powerless to affect the senses." Which is to say, if a metaphor is a kind of local transit in its Greek origins and function, as I was asserting in a natural history museum long ago, truth is a metaphor with a flat tire. New Age people are literalminded, wanting a vast number of contradictory things to be literally true, looking for an absolute point of origin or arrival which will bear an absolute truth, maybe the point at which they can stop traveling. Perhaps my difference with them is that New Age people want to arrive and I want to travel.

So I trod off through the town to the forest as often as my feet and the allure of information allowed. The town itself had medievally narrow streets, a population that seemed more in retreat from their fond invaders than most towns I saw, and crowds of carriage-drivers hovering hopefully near the center of town, their magnificent heavy horses shifting feet patiently while the drivers talked among themselves and hailed passersby. The forests of Killarney began right at the edge of town, behind a stone wall, and spread up into the heights of MacGillicuddy's Reeks, the highest mountains in Ireland. The trees immediately inside were huge, each one spreading its branches to carve out a great cone or sphere in the air. Winds made the treetops roar and wave their branches wildly, while the air I moved through below was still and calm.

By virtue of their verticality, trees resemble the other upright species, human beings; and so these had the stately presence of ancients and witnesses, presiding over the earth and sky they link with root and branch. It was impossible to regard these trees without thinking of their immobility through disasters and revolutions, the immobility of place rather than of truth. It isn't hard to imagine becoming a tree, as various mortals in the

Greek myths did, to imagine one's feet sinking into the ground and becoming fixed, one's arms spread out in a kind of benediction become solid enough not to suffer from gravity over centuries, and with these transformations comes a sense of overwhelming peace. The surrealist photographer Man Ray once reflected, after visiting the redwoods near San Francisco, "They are the oldest living things in nature, going back to Egyptian antiquity, their warm-colored, tender bark seems as soft as flesh. Their silence is more eloquent than the roaring torrents and Niagaras, than the reverberating thunder in the Grand Canyon, than the bursting of bombs; and is without menace. The gossiping leaves of the sequoias, one hundred yards above one's head, are too far away to be heard. I recalled a stroll in the Luxembourg Gardens during the first months of the outbreak of war, stopping under an old chestnut tree that had probably survived the French Revolution, a mere pygmy, wishing I could be transformed into a tree until peace came again."

Along the path to the ruins of Ross Castle on the lake, the trees became smaller and clustered together, more truly forestlike. In the boggy ground near the water's edge, birches or aspens had tipped over, their shallow circle of roots still holding onto the black earth, and where they had been uprooted, a pool of water formed. These pools were round too, and reflected the sky through the clearing that had been opened up by the trees' downfall, while the circle of dirt-clutching roots poised next to each of them suggested the black lid of a lady's mirrored compact. Daffodils, bluebells, yellow bearded irises, and purple rhododendrons grew in this forest, blurring the boundaries between the wild and the tame. The red deer I saw in a remoter meadow ran when they glimpsed me, not like the deer I know, but with a strange flattened gait that seemed more primordial. When they reached the barrier at the far end of the meadow, however, they wheeled back as though it had all been for pleasure and didn't seem truly wild at all. They must have been descendants of a herd kept for gentlemen's pleasure when the forest had been a hunting park, for a plaque in the Dublin Natural History Museum declared all red deer had been introduced or managed since the thirteenth century. I remembered an American wilderness advocate who said, Wilderness without wildlife is just scenery.

It was scenery that the scenic tourists had come to admire. Scenery is a forest with poetry instead of wolves, and European ideas of nature often derive from this dewolved and much deforested landscape. The Russian poet Joseph Brodsky wrote, " . . . when a European conceives of confronting nature, he walks out of his cottage or a little inn, filled with either friends or family, and goes for an evening stroll. If he encounters a tree, it's a tree made familiar by history, to which it's been a witness. A tree stands there, as it were, rustling with allusions. Pleased and somewhat pensive, our man, refreshed but unchanged by that encounter, returns to his inn or cottage. . . . Whereas when an American walks out of his house and encounters a tree, it is a meeting of equals. Man and tree face each other in their primal power, free of references: neither has a past, and as to whose future is greater, it's a tossup. Our man returns to his cabin in a state of bewilderment, to say the least, if not in actual shock or terror." Bewilderment or shock if he still expects the world to be Europe and nature to be scenery, at any rate; pleasure if he likes his emblems unstable, incompletely assimilated into culture. When I was younger, I used to envy Europe for having culture and long lines of history and tradition streaming back from every person and place. Now when I visit, theirs seems a place that poses too narrow, too domesticated a definition of what it means to be human.

Two of the Irish speakers had already leapt the gap between spiritual and political language with fine poetic talks, and it was interesting that of all the speakers they alone seemed to see Christianity as akin rather than opposed to the more exotic forms of spirituality being described. (Although pre-Christian Celtic culture is of great interest to overseas pagans and New Agers alike, neither group seems to make much of the country's last fifteen hundred years of spiritual practice, save to point out pagan survivals such as holy wells and the goddess who became St Bridget. Given that Ireland is an extraordinarily religious country and its religion is overwhelmingly Catholic, this is something of an oversight: the Commission on European Values had reported in 1984, "When it comes to belief in 'the soul,' in 'life after death,' in heaven, and in prayer, the Irish are so far ahead of the rest of the western world that any comparisons are totally irrelevant.") Dolores Whelan, who organizes opposition to the British nuclear power plant at Sellafield in England (in the Lake District of Wordsworth and Beatrix Potter), which routinely belches radioactive effluents into the Irish Sea, spoke about Celtic spirituality. She recited the mythological history of Ireland from the Fir Bolg and Tuatha Dé Danann to St Brigid and the present. "Heaven is a foot above your head," she said, speaking of the capacity to dwell in the imagination. And John O'Donohue, who was a priest and poet as well as a key player in the environmentalists' fight to protect the stony western expanse called the Burren from tourist development, gave a talk called "Stone as the Tabernacle of Memory" that elicited a near-riot of enthusiasm by its end, as though poetry's forever shifting imagery rather than prosy literal truth was what the audience really craved, whether they knew it or not. "Landscape is the firstborn of creation," he began. "It was here long, long before we were even dreamed. It watched us arrive. How strange we must have seemed: separated, single human strays wound around ourselves, belonging neither to the territories of our interiority nor to the outside territories of landscape. To the ancient eye of landscape, we must have seemed haunted." And he went on to speak of the silence of landscape, the memory of stones, and the presumption of people who imagine they can own land.

Towards the end of the conference, the Irish revolted. Several of the participants and audience members got up on the stage of the tent that had been set up for the largest events and read a manifesto – a very polite manifesto, I heard, though I could never find anyone with a copy. One Irish participant told me they'd thought the conference was held in Ireland because of an interest in their country's issues and culture and were annoyed to find that like most conferences it was taking place in the limbo of whatever slides were shown and subjects were discussed. The New Age wore out its welcome in Ireland then and there, he said. The locals took over the end of the conference, though, with a fife, drum, and fiddle and a long line dance that picked up everyone in the pavilion and led them out as a long chain snaking into the sunlight, where an old woman vigorously step-danced next to the musicians. When I left, a middleaged man sitting on a tree stump – another professional Irishman – buttonholed me and read me Gaelic poetry, which had the fine bristly sound unknown languages often have and whose letters lay round and inscrutable on the page of his old book. It was as good an answer as any.

10 And a Pound of Lead

There was an implicit analogy throughout the ITA conference between the two exotics brought in to flesh out or ground the transpersonal psychology, the Irish and the Native Americans, an analogy that was never articulated. It's one with an ancient lineage, dating back to the Tudor era, when the English were colonizing Ireland and, soon afterwards, the eastern edges of North America, and it's survived into the present. When the plantation of Ulster – that is, Ulster's violent colonization and conversion to a market economy – began, one of its executors, Fynes Moryson, wrote that the program should be carried out with "no less cautions . . . than if these new colonies were to be led to inhabit among the barbarous Indians." The historian Nicholas Canny comments, "It was ominous for both the Irish and the American Indians that authors frequently made cross-cultural references, thereby implying that the two were descended from the same primitive ancestors or that they were at the same retarded state of cultural development." In 1600, a mediocre poet wrote of the Irish kerns, or mercenaries:

> Fraught with all vice, replete with villainy
> They still rebel and that most treacherously.
> Like brutish Indians these wild Irish live;
> Their quiet neighbors they delight to grieve.

Some contemporaries compared the Irish instead to other exotic peoples, to Russians and Tartars, for example. The poet and colonial administrator Edmund Spenser went to great lengths to demonstrate the Irish were really barbaric Scythians; and an anonymous tract by a Munster landowner from the 1620s or 1630s sets out to demonstrate that Irish serfs were descended from demons, not humans. Many explorers of North America compared the peoples they encountered to the Irish for their material culture, for weapons, trousers; George Percy, walking near Jamestown, Virginia, in 1607, even compared a pathway he came across to an Irish bog or forest road.

Such comparisons were still being made two centuries later when Alexis de Tocqueville and his friend Gustave de Beaumont toured Ireland after Tocqueville's more celebrated investigation of democracy in America. "I defy you, my dear cousin," he wrote in a letter, "whatever efforts of the imagination you may make, to picture the misery of the population of this country. Every day we enter mud houses, covered with thatch, which do not contain a single piece of furniture, except a pot to cook the potatoes. I should have believed myself returned to the huts of my friends, the Iroquois, if I saw a hole made to allow the smoke to escape. Here the smoke goes out by the door, which gives, according to my weak lights, a decided advantage to the architecture of the Iroquois." By the middle of the nineteenth century and the height of the Famine, *The Times* of London could rejoice, "In a few more years, a Celtic Irishman will be as rare in Connemara as is the Red Indian on the shores of Manhattan"; the Irish were not only like Native Americans but would be displaced with the same disregard.

But by the end of that century, the values accorded the analogies had begun to shift. In the last decade of the nineteenth century, the great ethnographer James Mooney documented the Ghost Dance religion among the Lakota, northern Cheyenne, Arapaho, Comanche, and several other indigenous nations who embraced this apocalyptic cult; Mooney's own Irish nationalism is sometimes said to be what made him sympathetic to the struggles of these indigenous Americans. A few decades later, Roger Casement came back from the Putumayo rainforests of Peru to Connemara where, as one of his biographers puts it, "starvation and squalor caused an outbreak of typhus. Then the lot of the Indian and the Irish peasant seemed to him to be much the same. He christened the locality the 'Irish Putumayo' and wrote that 'The "white Indians" of Ireland are heavier on my heart than all the Indians of the rest of the earth.'" California historian Mike Davis tells me that when he first arrived in Belfast in the 1970s, he was sitting in a pub in a nationalist area when a drunken republican came over, slammed his fist on the table so that the glasses jumped, and demanded, "*Can ye tell me why ye killed Geronimo?*" The IRA has long identified with the guerrilla struggles of Native Americans trying to protect their homeland against colonialism; and the American Indian Movement activist and writer Ward Churchill reiterated the Irish-Indian analogy in 1994. I had first heard it a few years earlier from the Cheyenne-Arapaho artist Edgar Hachivi Heap of Birds, who'd spent several months in the Republic of Ireland. He came back and declared that as far as he was concerned Ireland was a formerly colonized nation of indigenous land-based people, the same terms in which Native Americans are often framed.

Though the analogy has been drawn for nearly four centuries, it isn't stable in its meanings. For the English colonizers, it was a way of pairing peoples whose lack of civilization justified their conquest; for Tocqueville and Casement it was a means to link two cases of material deprivation; for the IRA, Churchill, and perhaps Mooney it brings together resistance movements seeking self-determination and cultural survival. But the more I thought about the analogy, the more it seemed to suggest differences rather than similarities. On the face of it, there are hundreds of distinct Native American cultures with radically different beliefs and means of sustenance, from nomadic hunters to sedentary agriculturalists, at least as different as Sicilians and Laplanders, so there's nothing generic to analogize to the Celtic Irish. All analogies have limits; two things are alike up to the point of their differences; and what the analogy of these two cultures locates is difference from the idea of Europe. A vague idea of what it means to be attached to a land that has been invaded emerges from the analogy, but a very specific fracturing of the idea of what it means to be European also emerges. Politically, Ireland resembled what were labeled the new and third worlds, and culturally its traditional tribal society had more in common with colonized people in many places than with the mercantile urbanized nation-states of the colonizers. It's an analogy that suggests how contingent both definitions are, or all definitions are. One can say European or Native only conditionally, aware of all the disparate qualities they group together, all the similarities they divide.

The Irish themselves, like practically every Christian group that ever felt exiled or marginalized, had a long history of comparing themselves to the Jews in exile from Israel, which didn't necessarily make them sympathetic to actual Jews, but in Northern Ireland the Irish Catholics are now often compared to the Palestinians as disenfranchised locals (and the IRA and the PLO have supported each other, and for that matter the American Indian Movement has identified with both). Across the Atlantic, the term *white* too shifts around when one tries to locate the Irish in relationship to it. A whole different set of analogies arose around the Irish in the United States in the nineteenth century: they were compared to African-Americans with terms derogatory to both. "*Low-browed* and *savage, grovelling* and *bestial, lazy* and *wild, simian* and *sensual* – such were the adjectives used by many native-born Americans to describe the Catholic Irish 'race' in the years before the Civil War," writes the historian David Roediger. "The striking similarity of this litany of insults to the list of traits ascribed to antebellum Blacks hardly requires comment. Sometimes Black/Irish connections were made explicitly. . . . In short, it was by no means clear that the Irish were white." Ireland can be a divining rod with which to locate the points of seepage in the cut-and-dried terms of identity.

The Irish the Tudors wrote about were unlike modern Europeans, in both sympathetic and unsympathetic versions. The Elizabethan Edmund Campion described them thus: "The people are thus inclined: religious, frank, amorous, ireful, sufferable of pain infinite, very glorious, many sorcerers, excellent horsemen, delighted with wars, great alms-givers, passing in hospitality." Ireland was, in the early seventeenth century, a heavily forested island made up of petty kingdoms – one of Henry VIII's correspondents identifies more than sixty countries – without an overarching administrative hierarchy, a high king. Both the presence of forests and bogs and the absence of centralization of either government or population made the island's people particularly hard to subdue, and the war to subjugate them was a prolonged, brutal, fragmentary affair that never wholly succeeded. The Irish of Elizabethan times, like the Irish of such pre-Christian sagas as *The Tain*, were largely pastoralists, rather than agriculturalists, and both cattleraiding and cattleherding kept much of the population mobile. *The Tain* is, in a nutshell, about Queen Maeve's cattleraid and its fatal consequences, with a great deal of topographical detail about various parties' wanderings across Ulster and Connaught. The poet Seamus Heaney writes,

> The royal roads were cow paths.
> The queen mother hunkered on a stool
> and played the harpstrings of milk
> into a wooden pail

Reading the historical and mythological material about cattleraids and border skirmishes, meals of milk and cow's blood, one is reminded more of the Masai of East Africa or pastoralist desert nomads than Native Americans, who were often nomads but rarely herders.

Nomads are an affront to centralized administration and the idea of borders, and the Irish were no exception. The aristocrats and professionals had their own varieties of mobility, and even the peasant class took to the forests and hills with their cattle during the warmer half of the year. "I believe that seasonal nomadism," writes the folk historian E. Estyn Evans, "is an important and neglected aspect of Irish social history." Spenser, a colonial administrator in Ireland from 1580 to 1598, complained that the natives tended ". . . to keep their cattle and to live themselves the most part of the year in bollies [from the Gaelic *buaile*, which originally meant a temporary enclosure and came to mean a mobile gathering], pasturing upon the mountain and waste, wild places, and removing still to fresh land as they have depastured the former; the which appeareth plain to be the manner of the Scythians, . . . driving their cattle continually with them and feeding only on their milk and white meats. . . . But by this custom of bollying there grow in the meantime many great enormities, unto that commonwealth. For, first, if there be any outlaws or loose people (as they are never without some) which live on stealths and spoil, they are evermore succored and find relief only in those bollies being upon the waste places; whereas else they should be driven shortly to starve or to come down to the towns to steal relief. . . . Moreover, the people that live thus in these bollies grow thereby the more barbarous and live more licentiously . . . for there they think themselves half exempted from law and obedience, and having once tasted freedom do, like a steer that hath been long out of his yoke, grudge and repine ever after to come under rule again."

Besides the pastoralist agricultural peasantry, there were other professions entailing other kinds of mobility. Poets and lawgivers formed privileged classes who wandered freely across boundaries, as did physicians. Craftspeople too were often nomadic, because in a largely townless society they had to seek out their dispersed customers. In his fine portrait of Irish society in the Elizabethan era, David Beers Quinn writes, "The hereditary status of the learned class and the mobile character of some of its members and of certain ancillary craftsmen and specialist groups, appeared disruptive of the influence of the state and of the policy of stabilizing the population. The creation of small, stable, landowning units was difficult to resolve with extensive grazing rights and summer transhumance, which involved a certain undesirable diffusion of property rights." The picture that emerges seems a vestige of a heroic era vanished from the rest of Europe, with its nomadism, its powerful poets whose elaborate curses were greatly feared, its far more liberal customs regarding nudity, sexuality, marriage, and divorce (Engels, in his *Origins of the Family, Private Property and the State,* reports of this society: "The reasons that entitled a woman to a divorce without detriment to her rights at the settlement were of a very diverse nature: the man's foul breath was a sufficient reason"), and its very easygoing version of Christianity.

A new and permanent schism had opened up between the English and the Irish with the secession of Henry VIII and his subjects from the Catholic church; religious differences would make Ireland's invading and indigenous populations separate and irreconcilable in a way they never had been before. The English occupation of Ireland thereafter would take on the qualities of a true colonization, distinct from the invasive migrations culminating in near-assimilations that preceded it. It opened up a wound that is still bleeding. Campaigns of vast brutality were launched to subdue the population, again and again. Lord Grey, who succeeded Sir Henry Sidney (father of the poet), went on a two-year campaign beginning in 1580, whose tally was "1485 chief men and gentlemen slain, not accounting those of meaner sort, nor yet executions by law and killing of churls, which were innumerable." Spenser himself, who had come over as Grey's secretary, penned the lengthy treatise *A View on the Present State of Ireland,* which makes it clear not even such brutal wartime tactics worked and recommends instead ravaging the countryside and subduing the populace by starvation. He described the results of such a campaign in the southern quadrant of Munster, and his account has become famous for its brutal clarity: "Although there should none of them fall by the sword, nor be slain by the soldier . . . notwithstanding that the same was a most rich and plentiful country, full of corn and cattle, that ye would have thought they could have been able to stand long, yet ere one-year-and-a-half they were brought to so wonderful wretchedness, as that any stony heart would have rued the same. Out of every corner of the woods and glens they came creeping forth upon their hands, for their legs could not bear them. They looked like anatomies of death. They spake like ghosts crying out of their graves. They did eat the dead carrions, happy where they could find them, yea, and one another soon after, insomuch as the very carcasses they spared not to scrape out of their graves. And if they found a plot of watercresses or shamrocks, there they flocked as to a feast for the time, yet not able long to continue therewithal, that in short space there were none almost left, and a most populous and plentiful country suddenly void of man and beast."

Conquest of the populace by sword and starvation combined as strategies to conquer Ireland in the century after 1550, along with ancillary programs like outlawing the poets. The oaks were cut down to build ships and barrels and use as charcoal in metal smelting, and vast expanses of forest were cleared away. It was a fringe benefit that deforestation eliminated refuges for outlaws and rebels, though some Irish nationalists like to assert that this was the main reason – a Killarney forester told me that's what he was taught in school. The new landowners, with the recklessness of people who didn't see the place as their home or their future, denuded vast forests without replanting, permanently altering the Irish landscape. As the Dublin Natural History Museum mentions, by the end of the seventeenth century, when the clearcutting was so comprehensive that Ireland became a timber-importing nation, *squirrels* became extinct; they were reintroduced from England in the early nineteenth century. Afterward, both landowners and historians would blame the peasantry for Ireland's deforestation: it is a charge Young addresses, and lays at the landlords' feet.

The forests around Killarney give a faint taste of what the island must have been like when such stately old trees covered much of the landscape and the vistas of contemporary Ireland and some of its starkness must have been unknown. By 1901 Ireland had been reduced to 1 percent forest; it is still the most deforested country in the European Community, though replanting has officially brought its forests up to 6 percent of the landscape. Unfortunately, much of what are now counted as forests are dense tree plantations, often of

Douglas fir and other fastgrowing nonnative conifers, which do nothing to restore the original ecology or any kind of wilderness. They are literally impenetrable, so close together are the treetrunks with their dead lower limbs forming barriers, dead because the sun is completely screened out by a flat canopy of trees of identical age and species. Without birds, without animals, without an understory of grasses and smaller plants, without any of the other elements that make up a forest, they may be at last trees without a forest.

In *Ulysses* Joyce parodies the links between forests and nationalism through the person of the brutal Citizen: "Save them, says the citizen, the giant ash of Galway and the chieftain elm of Kildare with a fortyfoot bole and an acre of foliage. Save the trees of Ireland. . . ." From this sentimental tirade, the text slips into a parody of tree mythology with a society wedding between a forester and "Miss Fir Conifer of Pine Valley," attended by a veritable grove of trees – Mrs Rowan Green, Miss Virginia Creeper, and so forth, with the honeymoon of course planned for the Black Forest. Trees, or groves, were known to have played a role in Celtic religion, and much celebration and mystification of trees took place in Joyce's time as part of the Celtic revival. But for all Joyce's sarcasm, the deforestation of Ireland was a serious matter. Its beginnings in Tudor and Stuart England signify the onset of transnational export economies that enrich nonnative stakeholders at the expense of the indigenous population and environment; the iron smelting and timber and barrel export foreshadow resource extraction economies throughout the undeveloped world, from oil pumping in Ecuador to timber cutting in northern British Columbia. The trees gave way not only to Englishmen but to capitalism and the modern age, and with their clearance the Irish were being forced into the worldview meant by the term European.

Neither trees nor poets were completely eliminated, however, nor did the un-European mindset vanish entirely. Sir Henry Sidney had declared open season on poets in 1566: "whosoever could take a rhymer . . . should spoil him and have his goods," says an account by the contemporaneous historian Thomas Churchyard. After giving an account of poets who were robbed and beaten by some Englishmen, he concludes that the "rhymers swore to rhyme these gentlemen to death, but as yet, God be thanked, they have taken no hurt for punishing such disordered people." Because the Irish poets often composed work to praise and spur on a lord's military feats, they were seen as propagandists, and their work was sometimes translated into English rather as the CIA used to decode Soviet communiqués. Spenser himself had some poems translated out of sheer curiosity and grudgingly admitted they had literary if not moral merit. The man who appears to have been the last wandering Gaelic poet didn't survive into independence, but his memory did.

Lady Gregory, the Anglo-Irish friend of Yeats and nationalism, wrote in 1901 about encountering old women in the workhouse in Gort, County Clare. They were arguing about the relative merits of the poets they had known in their childhood, including Blind Raftery, who had died sixty years earlier, not long before the Famine. Another person, whose father had known Raftery, told her, "He was someway gifted, and people were afraid of him. I was often told by men that gave him a lift in their car when they overtook him now and again, that if he asked their name, they wouldn't give it, for fear he might put it in a song." He played a fiddle and wandered the roads of western Ireland, according to the accounts, a blind man who could compose songs, curses, and praises with equal facility, and his curses were feared as much as poets' curses had been in pre-Tudor times. One of Raftery's is said to have "withered up a bush." But, concluded Lady Gregory, "It is not easy to judge of the quality of Raftery's poems. Some of them have probably been lost altogether; some are written out in copy-books by peasants who had kept them in their memory, but, some of these books have been destroyed, and some have been taken to America by emigrants." So in 1901 a faint fringe of the forests remained, and a fading memory of the old poets, while a new crop of poets was beginning to serve the political purposes that had been part of pre-Tudor Irish tradition.

What is most peculiar about the war against the poets and trees in Tudor-era Ireland is the close involvement of the two greatest English poets of the age, Sir Philip Sidney and Edmund Spenser. Sidney was an aristocrat whose father – the one who declared open season on poets – had been Lord Deputy of Ireland. Sir Philip used to come over and spend summers with his parents, was involved in his father's political work, and later, when he became a courtier and diplomat for Elizabeth I, carried out missions in Ireland himself. Spenser, two years older than Sidney and the son of a ropemaker, rose in the world through a benevolent acquaintance and his own intelligence; he went over to Ireland in 1580 as a secretary, secured an estate at Kilcolman in northern County Cork, and immediately became unpopular with the neighbors. When Sir Walter Raleigh visited Spenser in 1589, the castle was surrounded with woods "of matchless height"; a few years later only bare fields surrounded the castle. Kilcolman had been the family seat of Desmond, and when rebellion broke out again in 1598, its usurper was a target. The castle was burnt, Spenser and his family barely escaped with their lives, and he died a few months later in London.

Or perhaps what is most peculiar about the war against the poets is that the English poetry of that time celebrates the pastoral. Most literally, the pastoral is about pastures – that is, about shepherds tending their flocks – and the pastoral poets' shepherd usually takes advantage of his leisurely wandering and watching to compose poetry. In the pastoral genre lies the origin of the modern aesthetic appreciation of landscape. Theoretically, then, a country of wandering poets and pastoralists should have enchanted the English rather than appalled them.

There's a continuing argument about when the Golden Age was. For a while in the 1980s, some rather New Age feminism located it in the agricultural matriarchies of the ancient Near East and portrayed these societies as humane, stable, balanced, sane, and possessed of a lot of other earnest virtues. The Men's Movement in this decade has often pinned the Golden Age on the more ancient hunting and gathering societies that are still far from vanished and has tried to relive it themselves – thus the ruckus over their appropriations of Native American cultural practices such as vision quests and drumming. A fine case can be made that agriculture is when finding something to eat became drudgery, because hunters, gatherers, and pastoral herders all have interesting, unpredictable, roving pursuits. After all tilling the soil is the sentence meted out to Adam and Eve when they're expelled from their own Golden Age in Eden. For much of European history, a highly formalized version has been located in a fictitious Arcadia, where the life of outdoor simplicity, wandering, leisure, lovemaking, and songmaking is summed up as *pastoral,* a word with the same root as *pasture.* The pastoral originated as a Greek genre in pre-Christian times when Theocritus located his goat-herd songs in Arcadia, one of the most backward, roughest parts of Greece; Virgil further refined the genre into an idyll of more intellectual shepherds and a more idealized Arcadia; and the genre throve in continental Europe before Sidney and Spenser made it English.

Sidney himself composed the first great English prose fiction, a fanciful romance called *The Countesse of Pembroke's Arcadia* which celebrates the elements of an aristocratic pastoral (he had also drafted, but never finished, a defense of his father's Irish regime titled *A Discourse on Irish Affairs*). Spenser rose to fame with his *Shepheardes Calendar,* a twelve-poem cycle of pastorals, and spent much of the rest of his life on *The Faerie Queene,* a masterpiece even more unwieldy than Sidney's *Arcadia* and as rife with celebrations of the archaic and the rustic. Fairyland and King Arthur were themselves, like many of the more specific elements of Spenser's hybrid poem, largely Celtic inventions, though *The Faerie Queene* also drew on classical mythology and Christian allegory. All these elements were put to work as part of a nationalist paean to Elizabeth herself, legitimizing the upstart Welsh Tudors as descendants of Arthur and symbolically marrying Elizabeth to her kingdom and its mythological past. In the *Arcadia* and the *Shepheardes Calendar,* Spenser and Sidney were laying the groundwork for the great English pastoral tradition, which has been central to that country's literature ever since, up through Thomas Hardy's *Far from the Madding Crowd* and Kenneth Grahame's *The Wind in the Willows* at the very least.

At the same time they were ensuring the impossibility of an Irish pastoral. (I realized during my trip that even walking tours were about an English, a Wordsworthian, idea of landscape and leisure, and had little to do with the relationship between people and land in Ireland; historically, Irish walks tend to be beggars' and nomads' circuits, flights and forced marches, displaced people's wanderings, all the unpastoral reasons and rhythms of movement; hillwalking, as mountain hikes and climbs are called, seems to be a relatively new minority occupation.) In 1994 the literary critic Terry Eagleton wondered at the absence of landscape from Irish poetry and concluded, "It would seem probable that a landscape traced through with the historical scars of famine, deprivation and dispossession can never present itself to human perception with quite the rococo charm of a Keats, the sublimity of a Wordsworth or the assured sense of proprietorship of an Austen." Even Milton's great pastoral elegy *Lycidas*, written a generation after Spenser, mourns for Edward King, who drowned in the Irish Sea on his way to join his father, John King, who was growing rich off the confiscated lands of Irish monasteries: the English pastoral seems often explicitly tied to the Irish antipastoral.

In paintings and poetry, the pastoral depicts a landscape where no clear boundary lies between nature and culture, the domesticated and the wild, between work and play; most typically, the shepherds pipe or converse in the shade of a tree while their flocks graze. Virgil imagined it as an explicit refuge from war and politics. Much as the Irish landscape was stripped of trees, so its natives were stripped of that margin of time and plenty in which the pastoral is possible. Irish poetry specializes in the antipastoral. Swift wrote a biting one in 1729, less than a century and a half after Spenser and Sidney had established the pastoral in English. Though his "A Pastoral Dialogue" owes something to his own bitterly antiromantic vision, it owes more to the Ireland around him. In it the nymph and swain who usually bear classical names are servants called Dermot and Sheelah, and they converse while uprooting weeds from between the stones of a courtyard, rather than while watching a flock. Literally tearing up the organic landscape with small knives, they protest their love for each other in terms that themselves reveal poverty and degradation:

My love to *Sheelah* is more firmly fixt
Than strongest Weeds that grow these Stones betwixt

says Dermot; and Sheelah finds a weed to use as her analogy too:

My love for gentle *Dermot* faster grows
Than yon tall Dock, that rises to thy Nose.

Poverty, sweat, lice, torn clothes, sharp stones, and liaisons with other partners occupy the rest of their pastoral dialogue. The pretty flowers and foliage of pastoral poetry have become weeds in this poem, just as they had become the shamrocks and watercress the starving Irish devoured, and the exported barrel staves and ship timbers. Pastoral imagery gone sour appears in another poem of Swift's written the same year, on the drying up of St Patrick's well, a poem about "the Pastors of thy rav'nous Breed/ Who come to fleece the Flocks, and not to feed," which features invasions by plagues of frogs, insects, vermin, and a tyrant "with his rav'nous Band," who "Drains all thy Lakes of Fish, of Fruits thy Land." The landscape is literally being devoured from abroad, and the harmonious abundance and benign stewardship of the pastoral have become the famine of the natives and the ravenousness of the rest, a world gone awry. Swift makes it clear that in damning the English one need not celebrate the Irish.

Forty years after Swift's antipastoral, another Anglo-Irish poet, Oliver Goldsmith, wrote a far more genteel and famous poem, *The Deserted Village.* It portrays what is widely believed to be his childhood village of Lissoy in County Longford, fictionalized as Auburn, "loveliest village of the plain," though it is usually used to illustrate English enclosure acts rather than Irish ones. Usually seen as a melancholic and mellow poem, it describes a traveler returned to the scene of his youth, where he had hoped to retire, and in places it is as stinging a lesson in economic consequences as Swift's poems. Both people and buildings have vanished, for the village has been uprooted to create sheep pastures for another landlord – a widespread practice in both England and Ireland at the time.

Ill fares the land, to hastening ills a prey,
Where wealth accumulates, and men decay

declares Goldsmith after delicate evocations of the vanished community, and soon thereafter,

Sweet Auburn! parent of the blissful hour,
Thy glades forlorn confess the tyrant's power.

He concludes with a prescient scene of uprooted peasants forced to emigrate to North America. Swift himself had condemned this practice of turning agricultural land back to grazing, eliminating "the Livelyhood of a Hundred

People" for a single grazier's benefit, and compares the graziers to Scythians. And with the comparison, we are back where the Irish pastoral started: with Spenser comparing the unconquered pastoralist Irish to the nomadic herder Scythians. This time around, however, the indigenous population had been weaned of the free-roaming life of pastoralists and turned into the poor tethered tillers of the fields (mostly potato fields) displaced by the very herds that had once been theirs, herds that no longer supported pastoral culture but foreign wealth. No longer mobile in the leisurely circuits of pastoralists, they would become mobile on the one-way routes of emigrants and exiles; like Goldsmith's traveler, they cannot return.

Even so otherwise lyrical a poet as Thomas Moore was moved to write the sarcastic "A Pastoral Ballad by John Bull," in which the five million bullets sent to Ireland in 1827 to suppress dissent are framed as a love gift from masculine England to a feminine Erin. Admittedly, a kind of pastoralism – the romance of the past and the peasant – comes from some of the Anglo-Irish poets of the Celtic revival such as Yeats and the later writers pursuing the official myth of noble peasants. Their sense of place and culture seems to have been a necessary underpinning for the new nationalism, but they evade the questions poets from Swift to the present pose, and their legacy lies more in postcards of "real Ireland" than contemporary poetry (though Yeats's "Ancestral Houses" of 1923, an uneasy poem about privilege, ends *The Penguin Book of English Pastoral Poetry*). The antipastoral continues through Irish literature, which excoriates the rustic life as thoroughly as English literature celebrates it, or at least describes it in clear-eyed unromantic terms – like Synge's turn-of-the-century plays about the harsh life of Aran islanders and Patrick Kavanagh's long poem about an isolated farmer, *The Great Hunger.* Written during the Second World War, Kavanagh's account of the dreary routine and unfulfilled desire of a bachelor farmer named Paddy Maguire is an epic of uneventfulness and spiritual famine. A long passage specifically mocks the romanticized peasant Yeats and tourists envision:

> The peasant has no worries;
> In his little lyrical fields
> He ploughs and sows;
> He eats fresh food,. . . .
> His heart is pure,
> His mind is clear,
> He can talk to God as Moses and Isaiah talked –
> The peasant who is only one remove from the beasts he drives.
> The travellers stop their cars to gape over the green bank into his fields: –

But his protagonist's mind is complicated, containing "The hysteria and the boredom of the enclosed nun of his thought" and the exclamation, " Oh Christ! I am locked in a stable with pigs and cows forever."

Ireland fractures the unity of Europe, the notion of whiteness, the Atlantic divide, and from it the cracks in the citadels of culture can be seen too. Spenser and Sidney, the poets of the pastoral, become the founders of an antipastoral, and the shadow of their political lives lies across their artistic merit. Not all English poets would be so compromised: Shelley, for example, came to Dublin as a boy of nineteen, long before his scenic tour of Killarney, and handed out tracts he wrote calling for an Irish revolution, a foolish but honorable act. Though the English poems were taught to me as milestones on the road to civilization, it is no longer clear which way they lead. The exquisite poetry of Spenser's masterpiece *The Faerie Queene* is inextricably linked to his brutal prose *A View on the Present State of Ireland.* Even one of his biographers refers to the two as a pair (though one was unavoidable, the other nonexistent, in my English literature classes). Should the magical trees he celebrated in the poem be weighed against the trees he uprooted in County Cork? Can one have the latter without the former, since Ireland's lack of a landscape tradition is rooted in its scarred landscape? Can one understand the presences of English literature without the absences of Irish literature? Are the presences in the former, at some level, bites taken out of the latter? Is England gardenlike because Ireland was prisonlike? Does the English pastoral, and the security and abundance it represents, depend on the impoverished land and people of other lands?

This is where the conflict between New Age and Native communities arose in recent years, over the depoliticized extraction of culture – though such depoliticizing views had already been established as literary scholarship. It was against this divide that poets became politicians in Ireland at the turn of the century, reviving language, folklore, mythology (including a romanticized version of peasantry and land), and the political function of poetry, watering the old roots and milking the old wounds.

11 The Circulation of the Blood

The giantess got off the train as I got off the bus in Ennis, in a nondescript station near a housing development, proudly announced by a pair of gateposts on which lifesize German shepherds sprawled, glazed in the pastel colors of china teacups. In other parts of the world, the animals might have been fierce, like Chinese lions or French griffins, but these guardian figures were almost asleep in a scene of stultifying peace. We were the only two stopping there, and she asked me in an American voice which way it was to town. The giantess had small slanting eyes and creamy skin and though she couldn't have been more than six feet tall, she was as massive as she was beautiful, with the kind of heroic scale that used to be called Junoesque. I felt quite a stick by her side as I led her where I suspected the town lay. We ended up traveling together for a little while, with that freemasonry of the road with which people fall in with each other and part far more casually than they do in fixed circumstances, where there might be consequences.

The giantess told me about herself, about her home in a part of our country as arid and expansive as Ireland was compressed and wet, about her marriage, her mother, and her grandmother. She had that sheltered quality fewer and fewer women will have, naive, tender, vague, like a sleepwalker who wandered through the world as though she were still cushioned in amniotic fluid. She was my age and, with the leisure her husband's income provided, still deciding what to do with her life. She had left a family party on the continent to come to Ireland by herself and sketch and muse about it. We found a place to stay, got rid of our bags, and walked more, through the dreamy town of Ennis, with its slow shallow river full of green water weeds and swans and its ruined church and the rest of its buildings set at odd angles seeming neither particularly new nor old, the town where, in *Ulysses*, Leopold Bloom's father committed suicide. For all her height the giantess was a slow walker, and she always trailed a step or two behind me.

As we passed along a tree-lined avenue, she told me about her Irish grandmother on a farm in the Midwest, who used tell her granddaughter about talking to trees, when the middle generation was not there to intervene. Their sense of time is different, said the grandmother, so that if you spoke to them or did something for them it might be a year or more before they'd say something, and maybe they wouldn't anyway. Over the blandest Chinese food I ever tasted, she told me about her mother, who was not so distant and romantic a figure, but a hovering source of anxiety: her mother wanted to live through her and never heard anything that contradicted her dreams of a daughter as a mirror of purity, self, and maternity. She kept making the giantess presents of things she wanted herself. And so we talked about mothers and the difficulties of daughterhood. Families up close, particularly the parent-child debacle, are often something one defines oneself against rather than with. But families at a distance, relatives with whom the ties of obligation and experience are not so tight, are another matter, a half-imagined community, and ancestry often seemed to me a wholly imagined one, a mythology of origins and membership – for it's not the dead ancestors but the surviving stories that provide sustenance. A myth strong enough to make the Irish government give me citizenship, strong enough to make the giantess and me come over to see how the place fit, but as often based in recitations and lists and names as in experience.

In the morning we both hopped aboard the same bus for the coast of Clare and the Cliffs of Moher, which the surfer who sold me my maps had told me to visit. On an impulse, I got off in the town of Lahinch, to be alone and walking again, and to let the giantess pursue her own plans of solitary adventure which she seemed in danger of giving up for the far easier approach of trailing along with me. The cliffs were a few miles away, but things seen to rise gradually out of their surroundings are infinitely more real than things which suddenly bump up in front of one, as though the theater set has been changed behind curtains. So I set off walking, along a sandy shore, and then through a network of streets that became roads and roads that became dirt lanes, still within sight of shore. There was a tower marked on the map, and when a few miles later I met a woman on a back road not far from the sea, I asked her how to get there.

She was old, she looked like my mother's mother with her straight back and blue eyes, and she was wearing a green sweater under a vivid green cardigan and watering the geraniums with a dirty milk jug in her walled front yard. A border collie and an unusually calm kitten stood by attentively. She told me to continue down the road and go through the gate and along the trail, even though it said bull in field – the bull never showed up, she said. That way I'd get to the first tower along the cliffs, even though – and I could never be sure she said this, but between the difference in our accents and the tenacious direction of her conversation I couldn't worm it out of her again – even though she'd never been there herself since she'd been here. I was flattened by the idea that she'd never bothered to walk all the way to the cliff's edge and wonder about it still. Could a life be so profoundly local, could daily life so overwhelm idle curiosity? And in a life so settled, what filled the place occupied by eternal adjustment for the rest of us? Was it more tranquility or more unconsciousness – deeper roots, shorter branches?

Did you move to this area, I asked, because she'd said *since she'd been here,* though her house looked too unpolished to be a holiday home. She said she'd been here her whole life. The house she grew up in was just over the rise, she indicated, in a tiny hamlet, and she'd married a boy she grew up with, had been married fifty years. Half a century seemed to merit congratulations, and so I offered them, and she replied that though it had been fifty years since she married, he'd been in his grave for the last seven of them. And they'd buried her brother a week ago. And if I saw two men quarrying slate by the cliffs, they'd be her son and nephew. She pointed at the wall across the road with her arm and said, It was here when I came fifty years ago and it'll be here fifty years from now. Then she asked me how I liked Ireland. Irish people were always asking this question, with the expectant confidence of a beauty queen inquiring if her hair was all right, and I said that I liked the unhurried cordiality of the people, though I can't have put it quite that way. Well we've not joined the rat race yet, she replied briskly, We're easygoing, like.

It's true that Ireland hasn't joined something that most often seems to be the industrial revolution, with its stern ideas about schedules and production. Time is not necessarily money here, where there aren't enough jobs to go around and there isn't a whole lot to rush toward or compete for (Joyce once remarked in a letter that the Irish are the most civilized people in Europe because they are the least bureaucratic). Or perhaps it's that with five thousand years of visible history breathing down your neck, any urgency is swallowed up in the vastness of time. Or it may be the metaphysics of Catholic countries – in which all time lies in the shadow of eternity, death is a more real and less terrible fact than in many places, and one's purpose on earth is not necessarily materially profitable – that makes Ireland so patient a place. It was certainly part of what made it a good place to travel through at the minimum speed of my legs, and an easy place to find people who would stop to talk. I wandered down to the pasture without the bull and found the tower immediately, a ruined square thing of huge stones a few yards away from the cliffs.

The celebrated Cliffs of Moher, which run for several miles along the southern side of Galway Bay, are a place where one can feel what it must have been like when Ireland was the westernmost part of the known earth. They still seem remote and lonely as they face into the vast beyond, an abrupt edge on the idea of Europe, maybe even lonelier now as the superseded end of the world. It was the kind of day when the sea rolls toward the sky and the sky to the sea and the exact point where they meet cannot be determined amid the bright blur of distance. A cloud like a puff of smoke floated over the slate blue of the three Aran Islands to the northwest, and their swelling profiles looked like whales, like the whales on whose back St Brendan the Navigator once celebrated mass. But the main thing was the sea, the inexhaustible sea whose waves rolled as regularly as breathing and which had been lapping the coasts since time and the pull of the moon began. The sea was a deeper blue than my own churning gray Pacific, blue as though different dreams had been dumped into it, blue as ink. I imagined filling a fountain pen with it and wondered what one would write with that ocean. As I walked along the cliffs I saw the rocky shoals at its feet where the water washed up a bright green. Two waves of green washed round a rock over and over, reaching for each other like a pair of foamy arms, and sometimes they joined on the far side and became a ring, and more green waves washed up on spiny reefs and ridges, then turned blue again as they returned to the depths.

Seagulls far below near the sea cried out in voices which rose up faint and eerie, and where the cliff curved in they seemed to fly out of the stone itself. The turf was so soft and pillowy it seemed to soothe one, as though all the world were so welcoming, and the primroses on their own beds of turf on the way down made it easy to forget what lay in store if one slipped from the path. The path itself proceeded often only a foot or less from the cliff's edge five hundred feet above the stony sea, bound on the other side by sunken pastures. Thin sheets of the local stone set up in tombstonelike vertical sections walled in the pastures, so that every field was funereal, the path a narrow ribbon between the slabs and the deep blue sea. A few miles away from the old woman in the green sweater I came across the quarry she mentioned, and when I looked down into it, a pair of men, presumably her son and nephew, were slowly gathering further slabs and sheets of this stone as gray as stormclouds. The slate loaded onto wooden dollies was cracking into squares and rectangles, as though already architecture. Rather as the sky shades into the sea here, so the made shades into the found, and the millennia of stone architecture seem as much an outgrowth of the stone all around it as an intentional erection.

The second tower along the cliffs, O'Brien's Tower, was far more crowded than the first, and a whole visitor-center complex with educational displays, snacks, and souvenirs had been built next to it. Linking all the structures were walkways paved in the local slabs. These slabs, however, had fossil tracks like worm trails winding across them, signs of a time that made the tower recent. Walking along them I saw the giantess in the distance, talking to a middleaged couple by the snack bar. It was too soon after we'd parted to have a reunion, and the bustle was disconcerting after all the quiet of the day, so I slipped away as the trail wound upward and the cliffs grew higher. The

American poet Wallace Stevens wrote about the cliffs, and about ancestry, though he wrote after seeing a photograph rather than visiting the original:

> . . . go to the cliffs of Moher rising out of the mist,
> Above the real
>
> Rising out of present time and place, above
> The wet, green grass
>
> This is not landscape, full of the somnambulations
> Of poetry
>
> And the sea. This is my father, or, maybe,
> It is as he was. . . .

Even for Stevens, "The Irish Cliffs of Moher" is an ambiguous poem, suggesting that landscape itself is an ancestor, the point of origin in place of ancestors, or that the generations die away but the cliffs remain: "My father's father, his father's father, his – / Shadows like winds." Or that when one goes looking for ancestors and antecedents, for time, one finds only enigmatic places.

After walking nearly fifteen miles that day my feet were beginning to murmur ominously again, so when I saw a bed and breakfast in a nondescript one-story house on the road into Doolin, I went in, handed over some cash to the sturdy middleaged woman whose place it was, and took a bath and a nap. Later, as it was getting dark, I walked into Doolin itself, along the main road, and then down to the one street that made the scattered buildings coalesce into a village, the street that runs down to the sea and the port for the Aran Islands. Doolin was once, not long ago, supposed to be a place where folk musicians came in the winter to teach each other tunes, particularly in a pub called O'Connor's. In pursuit of this authentic culture came tourists, and now though the two or three pubs have musicians in them, they're hired people playing for an audience rather than each other. The tourist brochures still insist it is a place where musicians come, though the next day a young German woman who had settled locally told me with scorn that everyone in the pub I was in probably spoke German and there was nothing special about the place anymore. But I did speak to a local in O'Connor's, a blackhaired young man with a nose like a snail shell who was back from Dublin, having a pint with a friend and enjoying the fiddler. He told me there's a special quality about west Clare and its music, and people from there have a special regard for each other when they meet elsewhere. We talked about music from various places, and he told me they were all for shutting down the pubs of Clare when a Seattle rock star – a genius! he exclaimed fervently – had killed himself a few months earlier. Then he turned back to his friend and picked up where he'd left off, checking in on a long list of local women who'd all married or emigrated and explaining his own plan to emigrate to Australia.

When I got up in the morning, I found that my fellow guests in the farmhouse were a midwestern American businessman and his late-adolescent son who'd gone to the next town over the night before, to Lisdoonvarna, because there were too many Americans in Doolin. Tourism, foreignness, is a virus many tourists suspect everyone but themselves of carrying, and just as a certain kind of traveler wants to be the first person ever to climb a mountain so another wants to be the only outsider in a pristine culture. This one, who left nothing out in his evocation of the ugly American, made me wish I was the only tourist at breakfast as he offered me unsolicited advice about what to do, told me stories whose climax came when he explained things to other people, shared the details of how much the trip was costing him, and confided his aspiration to buy back the family home of a few generations ago and retire out here.

The loud father and the silent son had spent the day before on the Aran Islands, a last outpost of authenticity, a place where, as he reminded me, they still speak Irish. He strengthened my resolve to avoid the place, whose population had been a kind of captive indigenous spectacle since the turn of the century, when J. M. Synge and various other folklorists and linguists would go there to learn the Irish language and marvel at the unspoiled primitiveness of the people. I'd forgotten how much I hate bed and breakfasts, because of the forced intimacy and because, as you try to slip past the family watching its television in the evening and inspect their decoration schemes as they serve you in the morning, staying in someone's private home seems like colonization wrought on the smallest, most genteel scale. The landlady of this place sat down with us and told us about how the Germans were buying up land in the vicinity for summer homes and complained of their unneighborly ways, while the American son looked down into his third bowl of cornflakes. I put on my pack and walked to Lisdoonvarna.

On my way there I wondered about the giantess who'd come because her grandmother had learned to talk to trees here, and wondered further what the midwesterner thought he would have when he bought his ancestral cottage, and what was the special quality of Clare the young man on his way to Sydney felt, thought of all of us circulating around the fixed point of an old woman who hadn't relocated any distance in more than seventy years. There are senses of ties that go back past the immediate family and home, to vaguer, more mythic things, to that ancestor in the people and the landscape Stevens spoke of. It's an odd business: the son of the midwesterner didn't even look like his father, though I look like the immigrant ancestor who died giving birth to my mother's mother. I saw her face in a tinted photograph for the first time a few years ago and have wondered since what else she may have transmitted down the generations besides the eyes I was looking at cliffs and tourists with. But that's a question leading into the unknown, a question about the extent to which identity is racial, the kind of identity my new passport endorsed.

The word *race* itself is thought to have come into English, French and German from an Arabic word meaning *origins, beginning,* and *head,* and the endeavor to identify oneself with race is in some sense a quest for origins, for a personal origin myth. There is a peculiar authority granted to origins in this culture, perhaps itself originating in the story of the Garden of Eden. The authority of origins asserts that in the beginning things were as they should be, and therefore everything afterwards is an unraveling, a decline, a sullying of original purity. Thus the true Irishman for my midwesterner would be not the young rock and roll aficionado in the pub briefly paying his respects to home before setting off for Sydney but some crusty old Gaelic-speaking fisherman on the Aran Islands ("it is worth adding, too, that the timeless Aran Islands of a J. M. Synge had a fishing industry directly linked by large trawlers to the London markets," writes Terry Eagleton). One could tell an anti-Eden myth in which it is destination rather than origin in which true identity lay: thus to be Irish is to be destined to emigrate, to love African-American music, to outmarry and mingle, and the true, ideal Doolin is only realized when one can hear several other languages in the pubs besides picturesque brogues and when the musicians there get paid. In such a myth, impurity and hybridity would be the ideal form all things aspired to, borders would exist only to be crossed, the urge to go backwards towards the origins before things moved around and mingled would vanish.

Both origin and arrival are unapproachable places, but race is an idea of belonging to something larger than the individual, with its origin and destination more remote, and vaguer. Behind every apparent origin lies another one, and origins nest within one another, each obliterating the one it succeeds. Race itself, this identification with an ethnicity also imagined as an origin, has for the last century tended to generate a kind of ethnic nationalism whose insistence on the inseparability of race and place is itself mystical. It imagines the nation as a single body, a body whose mystical unity is articulated in the image of a common blood, the blood that ties its members together as one race: "the dead generations from which she [Ireland embodied as a woman] receives her old tradition of nationhood," says the opening sentence of the proclamation that accompanied the 1916 Easter Rising. Before the emergence of ethnic nationalism, the unifying body was not an abstraction but that of the king, the divinely ordained body at the head of the feudal order. The two bodies of the king, the literal royal body and his ageless body politic, were replaced in the national imagination by the image in which the ethnicity or country constituted a single body in which all the chosen had membership, the nationalist body.

This is the image behind blood, the enigmatic fluid so often invoked in the ferocity of identitymaking, and blood insists on an unconscious, inherent identity rather than the conscious identity transmitted in stories and values (in other words, the nature/nurture muddle at its most political). "It is one thing to sing the beloved. Another alas," writes the German poet Rilke, "to invoke that hidden, guilty river-god of the blood." Blood signifies many things: the feminine blood of defloration and menstruation and childbirth, the sacral blood of Christ's sacrifice reiterated as communion wine. But the masculine blood of identity intensifies conflicts and alliances: blood feuds, blood pacts, blood brothers, blood guilt, baths, money, oaths (twenty zealots who took blood seriously signed the Ulster Covenant of 1912, against Irish home rule, in their own blood). This metaphorical or allegorical invocation of blood proposes that identity itself is inherent and transmissible.

Many governments particularly subscribe to this mysticism of blood now, notably the Israeli and the German. The Irish Republic now only grants citizenship to children and grandchildren of emigrants, but for Germany there is no statute of limitation. Ethnic Germans who settled in Russia three centuries ago have been repatriated in the reunified Germany, while second-generation German Turks – and all the generations of their descendants born in Germany – will never be eligible for citizenship under the current law. Israel itself was founded on the idea that the legacy of blood entitled the Jews to a legacy of land, an even more extreme version of the authority of origins. I've always been as much appalled as awestruck that a people could not only retain their distinctness during the nineteen centuries between the fall of Jerusalem and the founding of Israel, but could remain so attached to an absent place of origin that everyplace else could be framed as temporary exile, no matter how appealing, no matter how long they stayed. Becoming native is a process of forgetting and embracing where you are.

The linking of blood to ethnic nationalism found its strongest advocates in the rhetoricians of the Third Reich. ". . . who could ward off, / who could divert, the flood of origin inside him?" says Rilke in the same prescient poem, "... he waded down into more ancient blood, to ravines / where Horror lay, still glutted with his fathers." The metaphors that were the foundations of the Third Reich invoked a mystical Germany that was a single person of one blood, and blood and soil were the Nazi way of describing the allegiance of a people with a place, with ethnic nationalism. In this vision of a single body, a unified blood, the status of the Jews was almost inevitable. Hitler wrote, "How many diseases have their origin in the Jewish virus!. ... we shall regain our health only by eliminating the Jew." And he spent so much time talking about blood as though all Germans were one body and the Jew as though all Jews were one virus invading it that he seems to have come to believe his allegories were literal truth. The historians Michael Burleigh and Wolfgang Wippermann write that he "made no distinctions between German and foreign, rich and poor, liberal, conservative, socialist, or Zionist, religious or nonreligious, baptized or unbaptised Jews. In his eyes, there was only 'the Jew'." It's a fascinating potent metaphor, the nation as body and the people as blood, but Hitler, like the new ethnic nationalists in Europe and white supremacists in America, never examined his metaphors very carefully.

Blood's most significant quality is that it circulates, though the Western world only found this out with Harvey's experiments in 1628. Which is to say, despite the archaic idea that heredity and identity are situated there, literal blood is not so much like the reservoir or bank vault of the body as its interstate highway system or its rivers: blood is what mixes things up, imports and exports, keeps them moving. And blood itself is not a pure, a singular substance, but a compound, made up of red and white blood cells, T-cells, oxygen, hormones, and other internal messengers and regulators, wastes, nutrients, antibodies. A healthy bloodstream is a very mixed community, and an updated metaphor based on blood would have to be one of multiplicity and mobility. Blood and soil make an even less appropriate pair for grounding identity racially and spatially, since they are both zones of profound transformation. The ecohistorian Paul Shepard describes soil as "a skin, mediating the mineral and biological communities." Just as blood moves through the body importing and exporting diverse substances to the outside world, so worms and microbes course through the soil, aerating it, turning it over and transforming things at the ends of their lives – corpses, wastes and decay – into fresh soil where the cycle will begin again. Soil is a festival of corruption and reinvention, the alpha and omega of all corporeal things.

Hitler wanted it both ways, wanted the vitality of blood and soil with the unchanging solidity of what would have been truer metaphors, bone and stone – but bone and stone signify the permanence of death. A country can never be exactly like a body, for its borders are even more permeable than skin, and usually only island nations have much topographic reason for their national boundaries. Nationalism is to some extent a fantasy of making the skin impermeable enough to keep out what lies over the border, and conservatives tend to speak of the purity of the body and the nation in identical terms, with words like *alien* and *foreign* and *contaminant* applicable to both (the great hostility toward homosexuality that Wilde and Casement, among others, have encountered seems to come from an anxiety about the penetrability of the borders of the male body; as Irishmen they were already viruses in the English cultural stream). A local population might once have been but no longer is like the old metaphorical blood of the metaphorical national body, for it has been mingling more than ever in recent decades. Ireland's renewed mass emigration is often referred to as a hemorrhage, another metaphor that prizes containment – but perhaps the impulse of blood, to make the metaphor modern, is to circulate. A nation is like a body up to a point, but it's a metaphor that intentionally dead-ends in a dangerously simple picture.

Family trees and roots tend to impose some of the same fantasies. The shape of trees, the diversity of roots and branches, the unifying trunk, makes them a useful metaphor for many things. In writing about whiteness and racism, for example, the historian David Roediger says, "If, to use tempting older Marxist images, racism is a large, low-hanging branch of a tree that is rooted in class relations, we must constantly remind ourselves that the branch is not the same as the roots . . . and that the best way to shake the roots may at times be by grabbing the branch. Less botanical explanations . . . are in order." Trees provide a useful image of many ancestors or many descendants, though here too forgetting is as useful as remembering: ancient generations almost never culminate in the single descendant drawn on some family trees, nor does any family descend from a single illustrious ancestor. An accurate family tree going back a few centuries or so would more resemble a forest of interlaced branches or some other such tangle than the neat trees one gets; even for our own Irish ancestors, my uncle had to draw up several trees, a small grove, which pushed the unknown back several generations but hardly did away with it altogether. And to take the metaphor seriously is to open up new possibilities (a secular English Jew once said, in the historian Simon Schama's hearing, "Trees have roots. Jews have legs"). Trees must have either seeds that fly, like the winged seeds of maples, or roots that go back forever to a true origin, to the point before ethnic groups emerged, back to the mythological common ancestors in Eden or the biological ones in Africa. Which is to say, you can find yourself in origins, though you may have to choose an arbitrary point – a great-grandmother, an ancestral home, the Irish Cliffs of Moher – and beyond that you can lose yourself in origins.

Even Irish identity depends as much on forgetting as on remembering, on forgetting that the Celts were not always Irish and the Irish were not always Celts, on forgetting that for all the conservative, stubborn tradition, they have also changed drastically again and again since the days when they were pastoralist tribes. Science denies blood, the existence of real races in the species, and provides even more fluid and slippery origins, back to the bones of eastern Africa, then to the point where the distinction between humans and other species begins to blur. And if you go far back enough, biologists say, the internal pulse of the blood becomes the external tide washing over the most primitive creatures in the primal sea; blood and seawater still have the same salinity. The point of origin for human beings, however, is said to be the upright bipedalism – the walking – that grew out of the species' move to the plains or savannahs of Africa. I like walking as the first thing anthropologists regard as having made humans human, that point at which they leave the forests and stretch upright like trees themselves, a rootless tree reaching for the sky. Go back, go back further, go back to the origin of the species, and you find not a fixed point but a walker, or rather walkers, not necessarily all walking in the same direction. So one morning in June I walked to Lisdoonvarna.

12 Rock Collecting

I was on vacation. Everyone else was on holiday. I left my knapsack in the hotel-cum-hostel with pub in the center of Lisdoonvarna and walked to Ailwee Caverns. I went less for the sake of caves than for the route, through the barren windswept expanse of limestone on the southwestern lip of Galway Bay called the Burren, from an Irish word for a rocky place. It was a long walk, and it was raining. Up a crest, past one of the horrible tree plantations, which had picnic benches scattered around its impenetrable mass, on the apparent assumption that a forest is a scenic recreation site place no matter what. Over a slope where stone was breaking through everywhere and some of the fields were not really fields but pavements of warped, riddled, hollowed-out pale limestone, in which little pools of water gathered, sweeping toward the misty distance. Past the place the map said had a side road going to the Blessed Bush, where the imprint of St Brigid's knees can be seen, in stone that must be less complicated than the terrain I was wandering through, for such a slight impression to be noticeable.

Down Corkscrew Hill, apparently named for the zigzag road down its steep side, into a more inviting landscape, with trees along the stone walls and soil rather than stone as its primary surface. Past a field, where a primeval white horse, with thick legs and Roman nose and stiff short mane, came up to greet me, to a country hotel. Inelegant rain garb shed, into the hotel, a sort of poor man's stately home, with its tattered prints and old volumes from scattered sets. Tea on a sort of settee in the room with the bar, a table away from the only other guests there, a well-dressed English family, three generations of women who looked like unhappy dolls and men who looked as immobile as the furniture beneath them. Outside again, the hills looked like topographical maps, because they had eroded into ledges or sills as regular as elevation lines, but beyond them was the sea. The cave was like many caves – long sinuous corridors resembling bodily passages, the literal bowels of the earth – and the tour guide was a young man with a good memory but no flair for reciting from memory, like most cave guides, but it was pleasant to be out of the rain. Then I ran into the Giantess on the road. She was better at vacation than I was: she had stayed in a town much closer to the caves and walked much less and met a man in a pub in Doolin who looked like Mel Gibson, but virtuously declined him. I thought we might keep running into each other, bound up on some parallel track of chimera chasing, but I never saw her again.

The next day it was raining harder. I walked to Kilfenora and bought a packet of chips and a chocolate bar in a dusty store where strangers or women must have been infrequent sights, because the man who sold them stuck his head right up against the dusty window and goggled after me, his tongue balled up between his teeth in concentration or wonder. The church in Kilfenora – technically a cathedral whose bishop is the pope – was half in ruins, full of graves, and had on one wall a carved fourteenth-century Bishop making a sign of benediction with the same two-fingered gesture drivers now salute each other with as they pass. No one passed me, however; the Burren on this stormy day seemed like an abandoned landscape, like the surface of a planet whose inhabitants had all vanished an indeterminate time ago. No cars, almost no birds, and signs of cattle in the fields but no cows, just a flat rocky expanse to the horizon, scoured and gnawed by wind and rain – nothing but botany, geology, meteorology, and ruins. The wind was making the rain so horizontal it tickled my inner ear. There were pockets of water in the hollows of the limestone, and the slabs all fit together like pieces of an eroded puzzle, with long fracture lines making rows of rock. Even the whitethorn trees seemed lonely, each set at a distance from the next along the walls.

I had the dead for company at the eleventh-century church in the crossroads called Noughaval, another roofless stone structure being strangled by ivy, like a nervous system choking its bones. The surrounding cemetery's headstones amid the wet grass and nettles ranged from a past weathered into illegibility, tombstones become plain rock again, to the near present, with plastic flowers for remembrance, all ringed round by more stone walls. I had been walking a long time in the rain when I finally arrived at the great portal tomb of Poulnabrone. There were a few cars parked on the roadside and, despite the No Entry sign, figures wandering the stony field on which it reared up, a vast slab of stone held up high by a few uprights on a low mound of unshaped stones. The uprights had kept the slab balanced as a roof for four and a half thousand years, in defiance of gravity or celebration of balance. The roof hovered at a slight angle, so that it didn't echo the horizon but pointed beyond it, soaring. As I stepped across the treacherous footing of hollows, ridges, rows, and dips, I saw a figure in ink-blue clothes approaching from another angle. He reached the tomb at the same time I did, and because the rain was coming down harder, he invited me inside with a hospitable gesture. These are always where there is much wind and water, he said as we stood on the low mound under the slab, and added, I come from Brittany, where there are many such things.

He was at ease and set me at ease too by immediately assuming affinity, as though since we were sheltered by the same megalith we must care about the same things and could skip the preliminaries. The Bretons, he declared in his French-accented English, are blueeyed Celts, though he was darkeyed and blackhaired himself, and he asserted that Breton was the most widely spoken of the surviving Celtic languages, spoken by far more people than speak Scottish, or Welsh, or Irish – he spoke it himself. I tried out a passage of Rimbaud on him: "I have my ancestors' pale blue eyes . . . only I don't butter my hair," but he didn't recognize it, at least not in English. We wandered together over the rocky ground to the next field, talking of stones, Celts, fairies, pilgrimages, and old places. We had both been down the old pilgrimage trail of Santiago de Compostela, which begins in Paris, but he had gone the whole way, and on foot. He was more beautiful than Mel Gibson, but this is not a novel: his mother was waiting for him in the car, a cranky Colette in leather pants unimpressed by the wet pre-Celtic monuments he had brought her to see.

There were places in the Burren that had never been inhabited and hardly disturbed, and when the Office of Public Works tried to build an interpretive center for tourists in one of them a few years ago, it prompted one of the most heated environmental campaigns in Ireland, a campaign to leave the place undeveloped that was at least temporarily won. In the daytime it seemed possible to believe that human beings were rare, solitary creatures who existed largely to rearrange the stone according to slowpassing fashions into tombs, stone forts, churches, walls, and that there was no other scale of time but the eons of geological formation and erosion, the millennia of architectural styles, the decades of building, and the hourly shifts of clouds and wind and rain. Every place exists in two versions, as an exotic and a local. The exotic is a casual acquaintance who must win hearts through charm and beauty and sites of historical interest, but the local is made up of the accretion of individual memory and sustenance, the maternal landscape of uneventful routine. The Burren seemed to be an old local place that was becoming almost exclusively exotic (which is not to argue against the pleasures of promiscuity or for never leaving mother). The decline in population since the Famine has nowhere been more precipitous than in the west, and of all the places I visited, the Burren felt loneliest for its abandonment.

At night things were livelier. A group of Welsh people was staying in the place I was, and they were an energetic bunch. By day they bicycled, kayaked, and climbed, less out of any evident enjoyment than out of a dogged sense of propriety: these were their holidays, and this is how holidays are spent. But in the evening they drank and sang in the pub attached to our hostel and looked happier. The first night, there were hired local musicians too, and the proprietor's three young daughters came down and gave us a show of step dancing. They wore elaborate, stiff costumes with full short skirts, and the information circulated that the three had won many awards. Stiff and immobile from the spine on up, their grave faces suggesting their upper body knew not what the lower was doing, no matter how their skirts flipped up and their feet flew. I always think step dancing must be an elaborate allegory about conscious suppression and unconscious expression of erotic energy, with its impassive head and body and aggressive legs, but that's another story.

The following night, a Friday, the young English busker who'd been moping around the place drifted into the pub with his guitar. His first song was "Dirty Old Town," a song very popular as a description of Dublin, and his dirgelike monotone suited it well. But he flattened the next song into the same melancholia, and the next. The crowd of young Dubliners who'd driven straight to the pub for their bank holiday and surrounded me in my corner seat couldn't bear it; they rushed out to their car, came back with a pair of guitars and politely wrested the evening from his mournful grasp. They sang pop and rock songs with cheerful tunefulness and with lined notebooks full of lyrics and chords to keep them on track. When they weren't singing, they chattered and poured pints of Guinness down their throats at an impressive rate and kept quantities of cigarettes smoking in the ashtrays. The Welsh gang chimed in, requested songs, and joined the banter. One of them played an Irish drum – a bodhrán, it's called – and one had a theatrical baritone of awesome volume, and between songs they bantered with the Dubliners.

There's a word, *craic* in Irish, *crack* in English, to describe this lively conversation, in which jokes and insults and compliments and stories are fired back and forth in playful volleys. It impressed me that such talk was so highly valued it had a name, and impressed me more to meet people who made their own entertainment, rather than consuming someone else's. I bantered a little with them and fell into talking about rock and roll with the guitar-wielding Dubliner sitting next to me. Music was beginning to be as useful for socializing with the younger strangers I ran across as weather was with their elders. This man had a particular devotion to U2 and told me about the U2 concert he'd flown to New York to see.

U2 is itself obsessed with America, and its songs are as likely to be about Martin Luther King and the Fourth of July as about Ireland's Bloody Sunday; for them rock and roll is itself about America. Later on my travels I found a little one-night-a-week rockabilly club in Dublin where a collection of young people had formed a sort of cult of American pop culture. Europeans approach many genres of American popular music with a peculiar reverence;

it's as mysteriously, exotically perfect for them as though it were Ming dynasty porcelain. The French worship of jazz is well known, and much obscure country and rockabilly music of the 1950s is reissued on German and English labels with scholarly liner notes. The music invokes a fantasy America, pared down to twangs and heartaches and rhyme schemes, the perfect world of a perfected art form, and artists like the Australian Nick Cave and Ireland's U2 conjure up an iconic America over and over again, full of wild horses and wanted men unburdened by the banality of the familiar. Country music, which after all originated among poor rural white southerners in the process of being displaced, is the most popular genre in Ireland now, edging out rock and roll. When I was there even the prepaid telephone cards featured Garth Brooks, the corporate cowboy himself, Ireland's number one pop star.

But in the dimly lit bar on a Dublin back street, this pack of rockabilly aficionados felt like an obscure religious cult, druidic, secretive; the songs they knew and steps they'd mastered seemed like initiation rites or incantations summoning up another time and place. They were perfectly friendly to me, though I wasn't dressed for the occasion and can't jitterbug. I fell to talking about favorite fifties country songs with one of them, and when I said, Johnny Horton, "Honky-Tonk Hardwood Floor," he pumped my hand with the fervor with which Stanley must have greeted Livingstone, and I was one of the anointed anyway. A DJ walked me back through the dark streets after midnight, remarking in passing, The Irish are eighty percent drunk and twenty percent depressed. He was clearly among the 80 percent and slipped a little from his hipness as we parted on O'Connell Street, saying, God bless you.

Having seen U2 on their first American tour stood me in good stead in Lisdoonvarna, and having been to the place in the southern California desert their album *The Joshua Tree* was named after didn't hurt any either. Late that evening, when the singing had turned back into talking, a shaggy, grimy, spry old man, a leprechaun of sorts, wandered in out of the rain with a fiddle case clutched in one dirty hand. He showed off on the fiddle, playing jazzy, experimental introductions to his traditional ballads and jigs and mumbling through his long beard in a brogue so heavy it took me a while to notice the German accent around the edges. Finally, after the last call for drinks – no liquor can be poured in Irish pubs after 11:30 p.m. in the summer months and 11 p.m. the rest of the year – an old local quavered some American country ballads sublimely, with accompaniment by the Welsh drummer and the German fiddler. When he was done and my last whiskey was downed, I wandered off to bed while the guitarists serenaded me from the foot of the stairs with an ironic rendition of Led Zeppelin's "Stairway to Heaven."

The following night, in Galway, it was a young woman who had the guitar and handwritten notebook of songs – and a loaf of soda bread baked by her mother. She was from a small town not far away and had come down for a weekend holiday after breaking up with her fiancé, and she sang in a tiny, good voice "Take Me Home, Country Roads" and "Walking After Midnight." I'd caught a ride into Galway with a man who was trying to start a business supplementing the region's scanty bus service with a van shuttle. In practice, this meant that he picked me up in Lisdoonvarna, dropped by the youth hostel a dozen miles up the coast of Galway Bay, loitered there for midmorning tea and cookies with a couple of other young women, dropped them off somewhere, picked up his kids from school while still driving me about, and discoursed marvelously on local matters all the while. Efficiency is an unfriendly virtue, and no one I met in Ireland seemed afflicted with it.

Stone walls occupied us for the first stretch. They were first of all, said my driver, the easiest way to unstone a field, and the more stony the field the thicker the wall. There are supposed to be places where the walls are ten feet thick for this reason, and there was a field I saw where boulders too big to build with filled almost a quarter of the cleared field. I was becoming a connoisseur of stone walls: the low Cork walls in which verdure had almost overtaken the stone; the neat solid walls of places where the rock came in flattish shapes; the particularly charming walls in which the stacks of horizontal stones were crowned with a row of uprights that always recalled rows of books on a bookshelf; and the ungainly walls in places where the stone came in irregular shapes, walls as loose as lace, full of holes the light and wind came through. My driver told me that a lot of the walls had been built in Famine times, because the English didn't believe in giving something for nothing, and a lot of them were intentionally useless. He pointed out the lines of walls running straight up the rocky ridges of the Burren, walls in places too rough for large animals and too low to keep goats and sheep in, pointless walls built by starving people. He called the tracery of stonework evidence of torture.

Galway was full of evidence of history and music clubs, and I wandered through both. There were a lot of histories to choose from. There was a modern cathedral on an island, through which communicants wandered from the end of a mass, and tall priests in lace vestments strolled. There was a plaque to commemorate Christopher Columbus's stay in Galway:

On these shores
in 1477
Christopher Columbus
found sure sign
of lands beyond the atlantic

said the sign on the stone pillar on one of the many waterfronts of this coastal town on a river. It had been defaced with a little anticolonialist rhetoric by someone who wasn't so enthusiastic about the old world's early enterprises in the new world. Nora Barnacle Joyce's childhood home was open to the public on a little side street, and when I wandered in I was happy to find that her husband had played tourist too, long before he became a tourist industry himself. He apparently subscribed to the idea that one can understand history and personality through the tangible space in which events unfolded. There was a letter of his framed on the wall,

> 26.VIII.09
> My dear little runaway Nora
> I am writing this to you sitting at the kitchen table in your mother's house!!! She sang for me "The Lass of Augrim" [the song the dying lover sings in "The Dead"] but she does not like to sing the last verses in which the lovers exchange their tokens. I shall stay in Galway overnight.
>
> How strange life is, my own dear love! To think of my being here! I went round the house on Augustine Street where you lived with your grandmother and in the morning I am going to visit it pretending I want to buy it in order to see the room you slept in.

Nora Barnacle's family's house was a tiny box wedged between larger buildings, with one small room downstairs and another upstairs, no running water, no gas, and an open fireplace to cook on. In it she had lived with her parents, five sisters, and a brother until many of them were adult – though she had been farmed out sometimes to her grandmother and to a convent – and her mother, Mrs Annie Barnacle, had lived there until she died at the age of eighty-four in 1940. The Barnacle house was so small no one could have even sighed privately, and psychic as well as physical life must have been communal. (Ireland never did give me a very specific sense of ancestral identity, save for the reminder that those names on the genealogy were most often the names of poor people, the heirs to centuries of poverty, and the poverty in early Irish photographs, of furnitureless cabins and ragged, barefooted people, suggested what that poverty might mean. I remembered my own mother's admonition that people who fantasized about the past always thought they'd be aristocrats, but the great majority of us were descended from peasants and were we to be magically transported back would be peasants at best.)

The woman who took admission gave her set piece about the history of the house and, when I asked if you could feel a person's presence by being in the space they'd occupied, added that it was a very good place to write letters and that Stephen Joyce, the Joyces' grandson, had come by a few times and approved of what they were doing. A sweetvoiced ordinary-looking woman in her fifties, she gathered steam from there and went into the most marvelous soliloquy, circular and lyrical, chaotic and enthusiastic, one that would undoubtedly have made Joyce himself happy. All I could recall from it were the phrases: When we get to the end of time, we will see everything is connected. We will see the thread that runs through everything. Because if you go back very very very far – if you look far back enough at a family tree and such – it's all connected.

13 The War between the Birds and Trees

I didn't realize I was headed for a convent until I was on my way there. I had met Kathleen at the Killarney conference when she had tried to interest a speaker in what was happening to her local forest. The man whose talk had been about the importance of forests wasn't interested, but I was, and so she took me off to tell me more, and when she found out I was a writer – a professional witness – she invited me to come and see for myself and stay in her community, which seemed to be the hub of the environmental activism she was talking about. She had a sense of great urgency about her local environmental crises, she was fluent in the jargon of environmental activism, and she was wearing white running shoes, jeans, and a pastel sweatshirt. In her late thirties, she had chestnut hair and an appealing face and an air of delicate yet vigorous youth.

Where I'm from, communities that care about environmental issues usually mean collective households of young radicals. But when I called up Kathleen, she told me that Sister Phyllis and Sister Agnes were in Galway City and would take me back to Portumna, and I realized how far away I was and that I was headed farther away: to a convent. Portumna is a town on the west bank of Lough Derg, the big lake the Shannon swells into as it divides County Galway from County Offaly. Anywhere else, it could be called sleepy and small, but by Irish standards Portumna seemed average, with a main street of shops, with schools, churches, and with a ruined priory and castle being restored as national heritage-cum-tourist attractions. The Portumna Sisters of Mercy had once run a residential school to teach young women the domestic arts of farmwives, but the days when domesticity required such art were largely over, and the four sisters lived in corners of the handsome stone school building that must have once held a hundred. All except Sister Noreen had lived in San Francisco – practically everyone in Ireland seems to have passed through the Bay Area – and two of them still taught school as they had there. To welcome me, they opened a bottle of elderberry wine a local had brought by and slipped hot water bottles into my bed in one of the abandoned student cubicles. The enormous differences between their beliefs and mine never came up; it's easy not to talk about sexual morality.

Kathleen, a former teacher of eight-year-olds, was being funded by the diocese to work as an environmental activist, and she had a fine tradition behind her. Early medieval Irish monks had once written nature poetry that took such pleasure in the birds, berries, trees, and wolves around them that they seemed more like the Zen-monk painters and poets of Japan than like the more familiar world-denying ascetics elsewhere in Europe. St Columba, says an old book of the Irish saints, lived in a forest in Doire – modern-day Derry or Londonderry – and wrote a hymn "that shows there was nothing worse to him than the cutting of that oakwood: 'Though there is fear in me of death and of hell, I will not hide it that I have more fear of the sound of an ax over in Doire.'" St Coemgen was praying with arms outstretched one day when a blackbird laid an egg in his hand; with saintly strength he held out his hand until the egg hatched. This minority tradition whose most famous continental exponent is Saint Francis may always have existed, though the church is often spoken of as though it had been a monolith of nature-hating throughout its history (in such works as Lynn White's famous essay "The Historical Roots of Our Ecological Crisis," the first of a cascade of writings blaming Western dualism for environmental disaster). My own blanket dislike for the church had been slowly undermined by such phenomena as the US bishops' condemnation of nuclear war, liberation theology, and heroic Franciscan antinuclear activists. Kathleen told me many speak now of the "Greening of the Church," of a reconciliation between Christian and environmental dogma. A case can be made that ecological sensibility has a long Christian tradition here, and another that it's a new sensibility rewriting the past to establish itself as traditional. But whether or not an environmental activist nun in jeans was an anomaly, I was her guest.

Kathleen cared about all the subtle disasters going on around her, and she bewailed the southern Europeans who came to bag songbirds by the score, the Germans who were buying up vacation homes so voraciously that there were German-dominated villages on the Shannon, the decline of the local farmers, the local forests, and the local lough. She bewailed it from inside the house, over tea in the morning and tea in the evening, from inside the forest and from a scenic vantage point over the lowlands of east Galway, on our way to the school where her old students sang a song to me in Gaelic and asked me who I'd root for in the World Cup, and as we drove past a dump draining into a local stream, near which Travellers' piebald carthorses grazed. She modified it with her earnest enthusiasm for ecological visions and positive solutions. But as far as she was concerned, the whole world she'd grown up in was being destroyed, and she was apparently right. She brought me together with various locals who were likewise watching their world disappear from under them.

Fintan Muldoon, a handsome old farmer in a tweed cap and coat who helped organize the rural farmers of the west, dropped by on a cool afternoon. Sister Kathleen shut us up in an upstairs parlor with a turf fire and a tray of tea things until I was better educated about Irish agriculture – though not as well as I might have been, because we often failed to understand each other's accents. We managed, though, and he told me that the average farm in the west was less than forty acres, and some of those acres were usually boggy or stony or steep, so the farms were too small to be run on the scale of agribusinesses. The artificial agricultural economy of subsidies and quotas created by the European Economic Community was skewed in favor of larger farmers, and the Irish Farmers' Association wasn't doing anything much for the small farmer either. Back in the thirties, he said, a lot of farmers remained bachelors for economic reasons, and a lot of girls emigrated if they could. In the last twenty or thirty years, he added, a niece or nephew would generally take over the farm, but the standard of living wouldn't be high, and it has come to the point where the niece or nephew won't take a farm today. So the small farms and small farmers are dying out, and young women are still scarce in rural Ireland.

Muldoon and his fellow activist farmers had gone to the bishop in Clonfert and then to the archbishop of Tuam, and the archbishop had told them, I am the archbishop of a green desert, meaning that however nice western Ireland looked it was becoming alarmingly depopulated. The farmers didn't want to see their farms become vacation homes or join the ruined cottages of the Famine times, and the clergy had a stake in keeping the devout rural culture alive. They recognize that rampant emigration is sapping their congregations; posted prominently in the Sisters of Mercy Convent was a sign that read, "Emigration is not an act of God. It is not the will of God that the Irish should be emigrants, the migrant labor of the English-speaking world, and now the migrant labor of the developed European countries. . . ." So the bishops of Connaught came together to work with the farmers on the Rural Alliance to Develop the West. They may succeed in modifying change, but not in stopping it: the government estimates that forty thousand small farmers will give up by the end of the century. The government has less to say about what this means for a country whose culture has always been rural, even if the farm is what much of Ireland has been fleeing for a century and a half.

Kathleen was reviving the old convent school gardens to teach organic farming, mostly to women and the unemployed, on the chance that organic produce could provide an income for some of them. She herself remembered growing up on her family's farm, where they grew a variety of vegetables, kept cows and chickens, and were largely self-sufficient. But most contemporary farms in Ireland specialize in cows and sheep for the meat, wool, and milk markets, and even potatoes are often imported from other European countries. The subsidized and rigged economies appalled her: she told me stories of ships loaded with surplus butter anchored offshore and doing nothing, of farmers paid not to produce. Her father still farmed, and she took me out to meet her mother, who fed me ham sandwiches and tea in the room where the old turf-burning range still sat, though they cooked on an electric stove around the corner. Her mother was charming – practically everyone in Ireland is charming, at least to casual acquaintances – and joked that they ought to marry me off to Kathleen's bachelor brother, who still lived at home but wasn't going to take over the farm. And I realized all the farm people I'd met had been old, from Cork to Clare.

The primary local food product, aside from milk, beef and lamb, was frozen pizza. While I was visiting Sister Kathleen's community, the public meeting to discuss the Green Isle Pizza factory's solvents and other effluents being dumped into Lough Derg was held in a convent classroom. Some of the locals were in favor of expanding the pizza factory for the sake of jobs, and others pointed out that tourism was also a source of employment, and a dead lough wasn't going to be very attractive.

I could smell a faint aroma of pizza in the woods around Barney O'Reilly's house. Barney O'Reilly had taken early retirement from his job as a state forester, in part because he disliked the new forestry policies being carried out, and he showed me around Portumna Woods and Lough Derg. The woods, which he had once taken care of as the head of a crew of about a dozen, were neglected now, he said, and they were being managed more for profit than for the public good. Happy to have a witness to what he thought was going wrong, he went into a long digressive monologue studded with Latin botanical names while showing me stately trees on which trainees had been permitted to practice tree surgery, harvested conifer stands that left eyesores, stone walls from the old estate being left to crumble and other signs of neglect.

He insisted that such things would never happen in America, where the forests were protected, and I spluttered and thought of the hundreds of thousands of acres of primeval forest being chainsawed every year, but he was hard to interrupt. His America may have been safe and simple, but his picture of the changing west was gloomy. There was a mass exodus from rural areas, he said, and the small towns and shops would close down as the farmers went, and the locals would go to the cities or emigrate. The large farmers would buy up the small farms, and more and more of the west would be planted to subsidized forests, which create far fewer jobs than even livestock do. So the west would be reforested at last, but not as a triumph of nationalism, traditionalism, ruralism. The region would belong entirely to agribusiness, the state, and the tourist industry. It would exist entirely for outsiders.

I had thought from the passion she exhibited that Sister Kathleen was protecting a magnificent wilderness from destruction, but it was only the small forest park of a minor aristocrat's estate that had been transferred from private hands to state ownership. The forest had taken me aback, not because of what it was becoming but because of what it had always been: a mixed, managed, muddled wood. It had been the parklands of the estate, and one could have walked its perimeter in a couple of hours. It had some very pretty groves of beeches (which are not a native tree) and oaks (which are) and Monterey Cypress (which are Californian), but while wandering in what I thought was a forest I would suddenly find myself among trees that grew in rows or abutted the adjoining golf course. I realized that this was a place where Europeans were indigenous and my own criteria about nature and culture were meaningless, since nature or wilderness in America means the state of things before white people showed up or a condition in which human beings are a minority population. In such places, one can at least daydream of a primordial world, however inaccurately, but this was a historical world. The stone walls had the same authority of antiquity as the oldest trees and were not an imposition on them, and the place, whether tended or neglected, was nothing but a garden. The most natural-looking bits were what had been arranged to suit the estate owners' aesthetic of nature, and so even the trees were monuments to aristocratic tastes and land use patterns. The animals, the wildlife – fallow deer, badgers, shrews – were, like the decorative beasts in the borders of medieval manuscripts, well-loved, but hardly central to the story. And the story I had barged in on wasn't about the preservation of the wild but of the local.

The island Barney took us to was more dramatic. We set off in a little boat with an outboard motor across the smooth blue waters for the place where the cormorants nested. It was a tiny island, tearshaped, perhaps the size of a city block. Covered from shore to shore with leafless trees, its ground was thick not with leaves but with white feathers and guano. The stench was terrific. Every tree had a nest in it, and the cormorants kept up a constant screaming as they flew back and forth from lake to young. The trees were mostly Irish whitebeam, which are endangered, Barney explained, and the cormorants are endangered too, though they're flourishing here. They used to nest on the seacoast but their habitat there was being destroyed, and so they had flown all this way up the Shannon to find nesting sites. The seventy pairs here ten years ago had become four hundred. Weighing the merits of endangered birds against endangered trees had me stumped, but Barney was all for the trees. The birds would survive because they were mobile, but the trees were helpless.

*

George Santayana once wrote, "Has anyone ever considered the philosophy of travel? It might be worthwhile. What is life but a form of motion and a journey through a foreign world? Moreover locomotion – the privilege of animals – is perhaps the key to intelligence. The roots of vegetables (which Aristotle says are their mouths) attach them fatally to the ground, and they are condemned like leeches to suck up whatever sustenance may flow to them at the particular spot where they happen to be stuck." But there is more to be said for being local than Santayana allows. To be local is to merge into your world and become vulnerable as it is vulnerable; to be a traveler is to become the pared-back person I was beginning to recognize, free to invent and learn, but not to live in that local correspondence between memory and landscape. It may be that memory requires a locale and a community, the continuity of reminders in the landscape and people with a shared frame of reference.

The mobile and local are not necessarily oppositional. The migrants and nomads who follow circuits that tie them intimately to multiple locales are polylocal and must not be confused with drifters whose passage is linear, not circular. And change is not always akin to motion; motion is simply one way to keep pace with or outrun change – or stagnation. I like to claim to be on the side of the birds, but being a bird was a vacation for me. I haven't changed regions since I was five. Watching places over extended periods, I could see how radically they were being transformed, while most of the transient residents who passed through thought they were seeing stable neighborhoods, towns, cities, ecologies, economies, and climates. The mobile person sees the landscape she passes through as static, because she changes faster than it does, but the stationary person sees that everything around is changing. Had I not spoken to anyone in Portumna, I would have thought it was a sleepy town in which nothing was happening, rather than a place in which the past was being unraveled faster than people could bear.

The rocklike foundation for identity an ancestral land is supposed to be was dissolving before my eyes into a river of transformations. The longer I passed through the Ireland that both the Irish and the Irish-Americans seem to imagine as a solid foundation, the more it seemed instead to be made up of a continuous flow of discontinuities and accelerating movements, of colonizations and decolonizations, liberations, exiles, emigrations, invasions, economic pendulums, developments, abandonments, acculturations, simulations. Another generation would sweep away rural culture, modify its Catholicism, assimilate to the European Community and the global markets and continue to emigrate, and through the hole they widened, the world would come pouring in. I stared out the window of the bus that took me away from Portumna, on an afternoon when the very air was still, and saw all the animals were lying down in the green fields like beasts at a nativity, waiting.

14 Wild Goose Chase

In books, however, the Irish were always turning into birds. The old legends and romances are full of them, blessed, cursed, and chatty. There were the Children of Lir, whose stepmother turned them into swans but let the four of them keep their voices and sing enchantingly for the nine hundred years until St Patrick came and their enchantment broke. There was my favorite impacted myth, "The Destruction of Da Derga's Hostel," in which birds and people are so akin they breed and speak together. Birds with instructive messages show up in various saints' lives and miraculous voyage narratives. At the Paradise of Birds St Brendan sailed to, the birds were devout spirits with instructive messages, and birds, of course, often serve as messengers, from Noah's dove to the raven who delivered bread to the desert fathers St Anthony and St Paul the Hermit. But they are peculiarly abundant in Irish culture.

The ethnologist Artelia Court writes, "The Irish tradition that it is unlucky to kill or molest swans is rooted in the belief that swans are the incarnations of human souls, frequently those of nobility, a belief reflected in present-day swan protection laws," and in the twentieth century folklorists encountered Irish Travellers who believed that cranes and swans might be one's grandparents. It may be a linguistic coincidence that the 1607 departure of the Ulster rulers Hugh O'Neill and Rory O'Donnell is in English called the Flight of the Earls, but the Wild Geese, another set of military leaders and soldiers who went into exile in France at the latter end of that century, were unmistakably avian (and may have had their revenge in the IRA's flying columns of 1919). In this century, birds came to crowd Yeats's poetry, some of them Greek, some Gaelic, like the watchers in "Cuchulain Comforted" who "sang, but had nor human tunes nor words" and "had changed their throats and had the throats of birds." Some are all his own: aristocratic swans, various hawks, cocks, and peacocks, and the artificial bird singing in Byzantium. Joyce was as bird-mad, and his books have birds all over them.

The centerpiece in this literary aviary, however, is *The Frenzy of Sweeney*, an early medieval masterpiece about a man become bird that has been intensively revived and reworked in this century. *Buile Suibhne*, as it's titled in Irish, was written down sometime between the thirteenth and sixteenth century but is thought to have been composed by the ninth century. A sort of fairytale Hamlet, its alienation, suffering, and ambiguity have a peculiarly modern cast, though it seems to mix Christian and pre-Christian elements. Like the two creations of Eve in Genesis, there are two explanations given for the birdbrained madness and literal flight of King Sweeney. One cause is a kind of post-traumatic stress syndrome from the horrors of the Battle of Magh Rath, the other, a curse by a holy man whom Sweeney had offended, so that the tale portrays the tension between the early church and those it had not converted. The battle took place in 637 between the High King of Ireland and the northeastern kingdom of Dal Riada, a sort of Scottish outpost in what is now Northern Ireland's County Antrim; a Suibhne who apparently evolved into the literary Sweeney seems to have taken part in it. *The Frenzy of Sweeney* tells in verse and prose the wanderings of Sweeney after the battle and the clerical showdown, on through to his death in another churchyard.

Birds are a pleasant image for the poet, who after all is a singer of sorts; Sweeney is a darker version of the artist as bird. Seamus Heaney speaks of him as "a figure of the artist, displaced, guilty, assuaging himself by his utterance." Though the narrative of *The Frenzy* takes place in prose, the speeches are in verse, and most of them are Sweeney's, and most of Sweeney's are marvelous. His madness is ambiguous. Contemporary interpreters have tended to suggest that he merely thinks he can fly, but the tale itself suggests that there is a morphological component to his condition. Certainly he takes up perching in trees, leaps and flutters great distances, and switches over to a birdy diet of watercress and other wild greens. The 1913 translation by J. G. O'Keeffe includes the fine digressive footnote that old Norse literature describes a battlefield condition recalling that of the Vietnam vets said to have taken to the deep woods: "Cowardly men run wild and lose their wits from the dread and fear which seize them. And then they run into a wood away from other men, and there live like wild beasts. . . . And it is said of these men that when they have lived in the woods in that condition for twenty years, then feathers grow on their bodies as on birds . . . but the feathers are not so large that they may fly like birds. Yet their swiftness is said to be so great that other men cannot approach them. . . ." O'Keeffe suggests that the Norse account may be based on the Sweeney poem. Levitation, he points out, is also a common characteristic of medieval saints, and he adds, "Until quite recent times it was the general belief in Ireland that madmen were as light as feathers and could climb steeps and precipices." Anguish, exile, distrust, and bird behavior make up the madness of Sweeney, and only the turbulence of the narrative seems to suggest the mental confusion that makes up modern madness.

This is how the story goes. Sweeney at the height of his kingly powers is beautiful, belligerent, and antiecclesiastical. After his many assaults on Ronan – a follower murdered, a psalter tossed into a lake, a spear cast at Ronan himself – the holy man cursed him: "thou shalt be one with the birds" and added nakedness, "madness without respite," and an eye-for-an-eye doom of death by spear. "Turbulence and unsteadiness, restlessness, and unquiet filled him, likewise disgust with every place in which he used to be and desire for every place which he had not reached. . . . He went, like any bird of the air, in madness and imbecility." His flight is an odd business. Sometimes he skims along the earth, sometimes he makes prodigious leaps, sometimes he can rise up out of the trees, often he falls. It has the uneasy, unsteady quality of flying in dreams, of movement sustained by concentration and will rather than innate ability.

I have spent twenty or more years flying in my sleep with varying degrees of ease and anxiety, and of all the things that tempted me to claim a mystical blood relation to Ireland, nothing tempted me like its recurrent flying literature. Dreams of flight seem allegorical, perhaps about the dreamer's sense of isolation, the unreliable ability to soar in language and imagination, and the desire to escape (and perhaps the weightless horizontality of the sleeping body). But even the usually prescriptive Freud admits, "Dreams of flying or floating in the air (as a rule, pleasurably) require the most various interpretations." In French the word *voleur* means both flyer and thief; in Irish it seems as though flying and madness have a cultural if not an etymological association. But Sweeney flies another way, in the metaphorical modes of soaring poetry and flights of fancy: in his marvelous passages of nature poetry. For he has become a lover of green places, trees, and solitude, and a lyrical praiser of them, so that *Buile Suibhne* joins the Irish medieval tradition of nature panegyrics, as well as topographical obsession. Magic and tragic, it also recalls the more enchanting side stories of Arthurian romance: Lancelot going mad and becoming a wild man in the woods; Sir Gawain accepting the challenge of the Green Knight and lopping off his head, only to have the conversational severed head set up another rendezvous in the deep woods; all the lonely wandering among hermits and wild places of the Quest of the Holy Grail.

Sweeney goes to Glen Bolcain, "ever a place of delight to madmen," with its windgaps, beautiful wood, clean wells and springs, clear streams, sorrels, berries, wild garlic, and acorns. He endures much bodily suffering – cold, injury, hunger – and bemoans that and the regal identity he has lost. He wanders over Ireland and then England and adds exile to his litanies of complaints. His wife takes up with another man. A kinsman lures him out of the trees with the lie that all his family is dead and then shackles him and brings him back to his kingdom. In his old home, a hag comes and drives him back into madness and longing for his forest home. With a great praise poem of the trees and plants, he departs:

Longing for my little home
has come upon my senses –
the flocks in the plain,
the deer on the mountain.

Thou oak, bushy, leafy,
thou art high beyond trees;
O hazlet, little branching one,
O fragrance of hazel-nuts.

Exile has become his home, home an exile. Sweeney has become an incurable malcontent on top of everything else, a hybrid birdman who belongs in neither place, and the only companionship he finds is with another birdy madman, an Englishman named Alan who goes off to die in a waterfall after they have been together a year. So it goes, episode after episode of Sweeney's misery. He ends up hanging around another church, living on milk he drinks out of a milkmaid's footprint in dung. The milkmaid's husband in a jealous rage stabs Sweeney with the prophesied spear. Pierced through the left nipple and mortally wounded, Sweeney dies in the doorway of the church, promised that he will go to heaven.

What about all these birds and the Irish tendency to turn into them? They suggest an identification with the ethereal, almost disembodied singing sweetness of birds, creatures of air rather than land, a kind of angel complex for people who have an easier time being spiritual than physical (the holy ghost is often represented as a bird, not least in the blasphemous songs of Buck Mulligan). And they suggest of course the dichotomy of nesting and flying, of the stationary and the mobile, though it is the mobile that dominates. They suggest freedom and the desire to be so rootless one doesn't even touch the mud and mire of earth, a counterimage to all the nationalist soil of Yeatsian poetry, and maybe to nationalism, to community, even to humanity and all the other ties that bind (though Sweeney ends up a tamed bird drinking from a footprint in the pastoral dung). The ongoing imaginative engagement with flight can also be read as a reflex of lack of freedom. Observations and fantasies of birds and flying things proliferate in the jail letters and drawings of Countess Markievicz, who participated in the Easter Rising, and, half a century earlier, the Fenian leader Michael Davitt documented his imprisonment in the book *Letters to a Blackbird*. *Portrait of the Artist*'s protagonist declares, "When the soul of a man is born in this country, there are nets flung at it to hold it back from flight." Heaney, Sweeney's best interpreter, writes that "it is possible to read the work as a quarrel between free creative imagination and the constraints of religious, political, and domestic obligation."

Sweeney is the bird as emblem of exile, and much of his anguish is that of no longer belonging where he came from and not being able to become part of where he has ended up, the hybrid's, immigrant's, halfbreed's, exile's double identity in which each half cancels out the other: "disgust with every place in which he used to be and desire for every place which he had not reached." Exile, from the Flight of the Earls to that of revolutionaries, peasant immigrants on post-Famine coffin ships, and writers looking for a more tolerant, diverse society, will become a major Irish theme, and so will Sweeney. In the twentieth century he makes a stupendous and sustained reappearance, in many guises. The writer Fintan O'Toole declares, "If traditional myths had to be used, then they would not be the heroic, collective myths of Cuchulain or Fionn MacCumhail, or of Joyce's Homer but the myths of Sweeney, the maddened existential outsider who was returned to again and again in the eighties in the visual arts, in theater, in poetry. Myth could be used, not to counteract the sense of fracture and isolation but to reinforce it." But Sweeney's shadow stretches not just over the last decade, but over the entire century: in T. S. Eliot's recurrent Sweeney figure, in Seamus Heaney's free translation of *Buile Suibhne* and in the Sweeney poems of his *Station Island*, in translations and references by many other Irish poets, and in the motifs of Joyce, to name some of his more prominent scribes.

Eliot and Joyce, the conservative deserting wide-open America for Catholicism, England and tradition, and the ex-Catholic cosmopolitan fleeing Irish narrowness, make a neat pair of opposites, and between them they seem to have torn Sweeney in half. Eliot's Sweeney is the stage Irishman of nineteenth-century prejudice in the poems "Sweeney Erect," "Sweeney among the Nightingales," "Mr Eliot's Sunday Morning Service," *The Waste Land* and the play *Sweeney Agonistes*. The Oxford professor Nevill Coghill says he asked Eliot, "Who *is* Sweeney?" and Eliot replied, "I think of him as a man who in younger days was perhaps a professional pugilist, mildly successful; who then grew older and retired to keep a pub." Herbert Knust makes an earnest case that Eliot had in fact found the 1913 Dublin Texts Society translation of *The Frenzy of Sweeney* and that his brute is akin to the tragic king, but no evidence indicates that Eliot encountered O'Keeffe's fairly obscure work (and it's unlikely he could have resisted dropping learned hints about it if he had). Apparently unaware of the original Sweeney and the ironies he was wading into, Eliot seems to have chosen the name to convey a bestial, fleshy, proletarian vulgarity for a character who is man become not bird but ape.

"Apeneck Sweeney spreads his knees": lines of Eliot are permanently embedded in my memory. The slim yellow paperback of Eliot's selected poems was the first real poetry book I ever encountered, and throughout my adolescence it stood for what poetry was supposed to be, a tight knot of impersonal erudition and description. Though college broadened my reading, it never questioned Eliot's supremacy, and neither did I until I came back to the Sweeney poems. I was aghast that such arid snobbery and revulsion was my introduction to poetry, and this time around it seemed to overwhelm the art. In "Sweeney Erect," Eliot writes:

(The lengthened shadow of a man
  Is history, said Emerson
Who had not seen the silhouette
  Of Sweeney straddled in the sun.)

He recoils from Sweeney shaving, sitting, and consorting with prostitutes; in "Sweeney among the Nightingales," he manages to work in animosity towards a second character named Rachel Rabinovitch, an apparently Jewish prostitute, who "tears at the grapes with murderous paws." The nightingales of the poem's title are, in the London slang of the time, also prostitutes. In contrast to Sweeney and Rabinovitch (whose names diligent Eliot scholars read as suggesting "swine" and "ravenous bitch"), heroic civilization appears at the end of the poem, and it's classical Greek. The nightingales "sang within the bloody wood / When Agamemnon cried aloud" and the poem ends with their "liquid siftings" staining "the stiff dishonoured shroud," apparently like pigeon droppings. This early Eliot seems truly odd upon reinspection, writing mostly about what he dislikes, not with the splendid wrath of a Dante but with purselipped loathing, and it began to seem that the fisher king whose groin wound laid waste the Waste Land was enthroned Eliot himself. Like Spenser's *Faerie Queene*, Eliot's *Waste Land* manages to appropriate Arthurian romance, a largely Celtic body of literature, while despising actual Celts.

He wrote of *Ulysses* that "in manipulating a continuous parallel between contemporaneity and antiquity Mr. Joyce is pursing a . . . way of controlling, of ordering, of giving a shape and a significance to the immense panorama of futility and anarchy which is contemporary history." For Eliot, the classics were a yardstick with which to measure the shortcomings of the present; for Joyce, a rhythm by which to beat out the present's own resonant music. Oh well. Joyce is said to have thought that *The Waste Land* was a parody of *Ulysses.* He liked birds, bodies, and the plebeian, and did better by Jews and women too. His biographer, Richard Ellman, writes, "So complicated in his thought and in his prose, Joyce longed to sing; a dream of his youth was to be a bird, both in its song and in its flight." Harold Nicolson thought the writer resembled a bird and in his diary described Joyce as "some thin little bird, peeking, crooked, reserved, violent and timid. Little claw hands. So blind that he stares away from one at a tangent, like a very thin owl." Joyce was delighted by his wife's maiden name, Barnacle, which implies not the shellfish themselves, but the species of Irish goose that in legend hatched from barnacles; beginning as coastal creatures clinging to a rock and ending as long-distance flyers, they make a nice emblematic Irish bird. Seabirds and geese are frequent in his work, as are birds generally. "There are sixteen geese in *Ulysses*," writes Nora Barnacle's biographer, Brenda Maddox, "and seabirds in myriad forms all through the *Wake*."

Portrait of the Artist teems with birds, and its accretion of bird images has often been noted. They appear every once in a while at first, and then, well into the book, after the epiphany of the wading girl who "seemed like one whom magic had changed into the likeness of a strange and beautiful seabird," real, poetic, and figurative birds begin to appear every few pages. There's a sort of crescendo of bird epiphanies, when Stephen stands on the steps of the National Library watching the flight of evening birds and wondering "What birds were they?" He contemplates them "building ever an unlasting home under the eaves of men's houses and ever leaving the homes they had built to wander." Four lines about birds from a Yeats play drop in, and then our hero kindly interprets the textual birds for us: "Symbol of departure or of loneliness?" Both, naturally: departure and loneliness keep each other company.

There's an actual Sweeney in *Ulysses*: the chemist Sweny who sells Bloom the lemon soap that crops up throughout the narrative, and there's a hint of magic about the alchemical chemist. *Ulysses* serves as a bible or encyclopedia of Ireland, though, in that everything can be found in it, from Parnell and Casement to geese and half the streets of Dublin, to say nothing of the occasional matzoh, smutty novel, and tinned-meat advertisement. The simplest temperamental distinction between Eliot and Joyce lies in the latter's ability to invest a cuckolded, kidneyeating, bathing, masturbating, brothel-visiting Irish Jew with a tenderly envisioned humanity. Bloom is, if not a Sweeney, at least a wanderer with something of the Wandering Jew about him, as well as Joyce's updated Odysseus. Like Sweeney, the Wandering Jew was cursed – in the latter's case, for failing to succour Christ as he dragged his cross to the Crucifixion – and like Sweeney he was an ambiguous figure, enhanced by his curse. Sweeney, of course, gains the ability to fly; the pedestrian Wandering Jew, immortality. He must walk the earth until Christ's return, when he will be forgiven and go to heaven, and so he too is an exile and a restless wanderer fated to die penitent in – since this is a Christian myth – the church's bosom. And what Sweeney has been to twentieth-century poetry, the Wandering Jew was to the nineteenth: a romantic image of the alienated artist.

But Bloom is not particularly cursed, except by his fellow man, and he's more everyman than artist. It's Stephen Dedalus who is not just a bird but a Sweeney, and Joyce gives him back his flight. Cunning, exile, and silence are the famous trinity he enumerates, but anguish, exile, and bird behavior suit him equally well. He is, after all, named after Daedalus, the Greek inventor of both the labyrinth and of the wings that took him and his son out of that labyrinth, and he is as anticlerical as Sweeney. In a particularly purple passage of the *Portrait,* he speculates on his name, seeing in it "a hawk-like man flying sunward above the sea, a prophecy of the end he had been born to serve and been following through the mists of childhood and boyhood, a symbol of the artist forging anew in his workshop out of the sluggish matter of the earth a new soaring impalpable imperishable being." Stephen Dedalus, not Leopold Bloom, because Bloom is at home in his alienation and in the perpetual exile of the Jews, and like Odysseus he is on a round trip that begins and ends at home. The possibility of making a home of exile – exile as a stable, familiar condition with landmarks and associations – emerges from Bloom in Dublin and Joyce out of it. It is Dedalus who is slated for an exile that shows no sign of a return, who represents his author and artists generally, and who rejects Bloom's offer of his home at the end of *Ulysses.*

Ulysses is a narrative of homecoming, of return from exile, adapted by a perpetual exile as a narrative of narrower scope: Bloom roams not the whole Mediterranean, but the labyrinthine streets of 1904 Dublin and ends at home, or at least a domestic semblance of it; Stephen Dedalus, however, will fly from Dublin for the continent, and for greater freedom. *The Frenzy of Sweeney* traces the whole route of the exile, from expulsion to penitential return, but it is the fluttering, accursed birdman who can neither stay nor go that is its lasting image.

Sweeney makes an intermediate appearance in 1939 in Flann O'Brien's first comic novel, *At Swim-Two-Birds,* where narrative, narrators, and characters in subsidiary narrations all mill about and argue. Considerable portions of *The Frenzy* that O'Brien translated himself appear with many chatty interruptions; and the novel itself is named after a place in the earlier work. Sweeney shows up in various other guises: in a wordless play by Ireland's Macnes theater company; in a performance titled *Revolted by the thought of unknown places . . . Sweeney Astray* in Amsterdam by the performance artist Joan Jonas; in translated snippets by various poets. The film director and writer Neil Jordan takes up the fascination with flight in his novella *The Dream of a Beast*: "Forget wings, he told me. Watch! He moved both arms as if stroking the air, stepped off the parapet and plummeted like a dead weight. I cried out in alarm, but saw his fall, of a sudden, transform into a graceful curve. . . . Wings are quite useless, he said . . . all one needs to fly with is desire."

In 1983, the same year that Jordan published this tale of metamorphosis from the human, Seamus Heaney reintroduced Sweeney with a free translation of the whole *Buile Suibhne*, and a year later organized much of his book of poems *Station Island* around the tale and its motifs of journeys, exile, and birds. Heaney calls his translation "a version" to license its freedoms of interpretation and a few excisions, and he titles it *Sweeney Astray*. With it he interjected Sweeney into the modern world to an extent neither O'Keeffe's translation nor O'Brien's version could; and with it he was not only reviving the Irish birdman for Irish literature, but redressing Eliot's swinish Sweeney. References to Eliot appear frequently in Heaney's prose, often to register disagreement or disapproval in such matters as the interpretation of Dante; he could not be unaware of the coincidence of Sweeneys. Eliot's apenecked former boxer is an antipoet; but in *Station Island* Heaney becomes Sweeney.

The majority of the poems in *Station Island* through its three sections concern journeys – flying and fleeing, walking, pilgrimage, driving a car, wandering, following – and they layer together various predecessors in the literature of journeying until all the layers resonate together. Birds appear throughout, united with the motifs of travel and alienation by the metaphors Sweeney supplies. The poem begins with an anecdote about the poet exiting from the London Underground with his wife, an uneventful emergence that becomes Orpheus emerging from Hades and Hansel "retracing his route," and such accretions of patterns persist throughout the poems. The central section, a prose explanation says, "is a sequence of dream encounters with familiar ghosts, set on Station Island on Lough Derg in Co. Donegal," an island also known as St Patrick's Purgatory. The first ghost is a Traveller – an Irish nomad – from the poet's childhood named Simon Sweeney, who appears with a lyrelike saw and advises the narrator to "Stay clear of all processions!" The introduction to his *Sweeney Astray* recalls that "the green spirit of the hedges embodied in Sweeney had first been embodied for me in the persons of a family of tinkers, also called Sweeney, who used to camp in the ditchbacks along the road to the first school I attended" – another restatement of who Sweeney could be.

The last ghost is none other than Joyce himself, "his voice eddying with the vowels of all rivers." The encounter is told in the terza rima of Dante's *Divine Comedy*, and as Dante made a Virgil to guide him where he wished, so Heaney creates a ghost Joyce to advise him. After all the reproachful ghosts of priests and victims of northern violence who precede him, Joyce advises about the poet's responsibilities and the Irish relationship to English: "Let go, fly, forget." And when Joyce speaks in Dantean meter, the masters – Homer and Virgil and Dante and Joyce, poets of wandering, seeking, and exile – fall in like a long line of travelers walking in each other's footprints, with Sweeney flying overhead, ready to guide the third section into further motifs of what one poem calls "a migrant solitude."

15 Grace

Sometimes it seems to me that time and memory are laid out in a secret geography that can never be mapped directly. Ireland was my third extended trip in a year, and each new landscape called up a corresponding set of dreams. By traveling across the surface of the earth, it seemed, one could begin to explore the geographies of memory, and what had been lost to consciousness could be recovered through judicious arrival in new places. In this sense, one can time-travel through conventional spatial movement, but not according to any comprehensible system. The dream map is not systematic, or maybe the dream territory is unmapped, or unmappable, but surely there have been travelers who traveled solely for the sake of the dreams to be found on strange pillows in strange lands.

In the Canadian Rockies the previous summer, a friend and a dog kept paying visits in my sleep as though their violent deaths had never happened, and the Rockies took on a strangely happy nostalgia; but that winter, I had familial nightmares in Guatemala, and it was hard afterwards to tell how much the dreams had made the place seem so ominous or if the place had made the dreams curdle instead. If every place had its resident dreams, then late-springtime Ireland was infested with former suitors, lovers, and other men, some of whom returned clear as yesterday though I had otherwise half-forgotten them. Some appeared as they seemed when I had been fondest of them, as though intervening history had been erased, and one laid all my books out in the horse pasture behind my childhood home, grass tufts surrounding each rectangle. My journey was, I said to him in a postcard, a very crowded solitude, but on this last leg of my wandering it was women who were beginning to fill up the days, women and the slippage of time.

It's part of the slippage in time of the traveler, this dream geography, as is one's attempt to rouse the past that gives the place its meaning. My friend Tim O'Toole visited a friend in Wicklow whose mother told him about Irish time when she was growing up. There were three versions of time, to be exact, rural, country, and official, all superimposed upon one other. Different people and different places subscribed to different versions of what time it was, and in order to be in synchronicity with them, to arrive on time, one had to know where their allegiance lay. In Tim's story, which of the trinity of times people chose seemed to reflect the past they were tied to: whether they ran by the old time or accepted a recent or modern idea of how the clock ought to be set. Time itself wasn't a truth but a political position, a stance in relation to the past.

Dream time, clock time, historical time. Being in Ireland always seemed like being in some past point of time, sometimes a decade or a half-century ago when it came to oldfashioned customs and unstreamlined ways of doing things, the length of memory and lack of hurry, the occasional horse-drawn cart in Dublin. I met an Irishwoman named Bride who told me her family felt snubbed that she'd left home without a good solid reason like marriage, and I thought of my own mother facing the same standard forty years earlier, a standard that had faded away in my part of the world. Bride, who had come to take a summer job in a tourist shop in Westport, was the one who casually remarked to me, Your country can't imagine its future because it can't remember its past, an idea it had taken me years to arrive at. But Ireland's future wasn't much more imaginable either, and the past came washing over it in torrents, along with my own.

It was the existence of the hermitess that seemed to throw me into the deepest past, however. Hermits! St Anthony was a hermit until his death around A.D. 355, and he and his fellow hermits make frequent appearances in medieval painting and literature. Fictional hermits were strategically placed in deep forests (no doubt cut down since) in Arthurian romances, where they succored imperiled maidens and counciled questing knights. Actual hermits and very small monastic communities had flourished in remote Irish places like the Skellig Rocks in early medieval Ireland, and some of the beehive-shaped stone buildings in which they lived and prayed are still standing. I had thought hermits died out in the Middle Ages, because by the eighteenth century German and English aristocrats who built hermitages in their landscape gardens had to hire hermit impersonators to reside in them and had a hard time keeping them to their ascetic life. I am afraid I grouped hermits with Irish Elk, as marvelous creatures who would never walk this earth again. But Sister Phyllis in Portumna had said to me, If you go to Westport, you should visit the hermitess, Sister Irene. She's a wonderful person. And the Sisters of Mercy in their big convent in Westport smiled upon me and told me how to get there.

Westport is a midsized town nestled on the east side of Clew Bay, about two-thirds of the way up the west coast of Ireland, in County Mayo. Clew Bay itself is squarish, and an archipelago of tiny islands clusters near its eastern side. The region is famous for two things: the mountain on which St Patrick is said to have fasted, and the pirate queen Grace O'Malley, who often operated out of Clare Island in the mouth of the bay. St Patrick's mountain was earlier called Croagh Egli or Cruchain Aigle, the mountain of Eagles, but it is now Croagh Patrick, a rocky coastal hummock of 2510 feet.

Patrick makes a good national saint for Ireland: born in England, or Scotland, or Wales, depending on whose version one takes up, he was kidnapped from a Roman or Romanized patrician family as a young man, discovered his faith while a slave in Ireland, and was granted miraculous foresight of the boat that would take him back out of Ireland, the first of many sea voyages Irish saints would take. So he was a multiple emigrant, or maybe just a migrant, and wandered western Europe for many years before taking up his missionary vocation in Ireland.

In Ireland itself he must have kept up his migratory ways, for sites all over the country are associated with him. He founded a missionary tradition that still continues, and he fasted on Croagh Egli/Patrick in imitation of Christ in the desert, Moses on the mountain, and various other moments of religious withdrawal into the wilderness. A bishop, Tirichan by name, wrote around A.D. 670, "And Patrick went forth to the summit of the mountain, over Crochan Aigli, and he stayed there forty days and forty nights. And mighty birds were around him, so that he could not see the face of the sky or earth or sea." Sometimes it's said to be the mount from which he swept all Ireland's reptiles into the sea. In recent centuries, the mountain has been the site of a devout pilgrimage during the last Sunday of July; I had seen photographs of middleaged pilgrims going devoutly barefoot over the loose rocks and scree of the chilly heights.

I had been walking down my own street when my friend Dana and I first told each other about Grace O'Malley. We opened our mouths to speak at the same time and found we were talking about the same woman. I had just come across a brief mention of her in a general Irish history, but Dana had a more lavish offering: Alice O'Malley, a young lesbian descendant who told stories about her piratical ancestress Grace. And so on another expedition I strolled over to Alice's apartment in the Lower East Side of Manhattan, in the old immigrant quarter of the city, among such streets as Hester Street, whose names I knew before I'd ever walked them. It was a cloudy Saturday morning, and the atmosphere of animosity New York's bustling crowds induce was absent. The city was deserted. The ten o'clock streets belonged to grime, to odd smells, and to me.

Between what Dana had told me and what I'd picked up about Grace O'Malley, I'd expected Alice to be imposing, but a small, narrow woman with cropped strawberry blonde hair under a plaid cap came downstairs. She gave me the impression of having met more trouble than joy in her life, a serious, intense woman, but hardly fragile. We drifted over to Alphabet City and ordered our breakfast in an Irish music club that had lost its liquor license and begun serving food. As a couple of families set up a children's birthday party in the middle of the room, Alice told me stories over our soggy eggs and damp potatoes.

The morning I set out, the mountain was covered not in birds but in fog, and my route pleased me: to reach the hermitess, I would have to circumnavigate the mountain, and so I would see all its visible lower faces and much of the scrubby and sparsely inhabited surrounding countryside. It was eleven miles clockwise from Westport to the Drummin junction where the road turns off for the hermitess's home. I began to worry I might not make her visiting hours, and so a few miles later when a car passed I stuck out my thumb. The middleaged farmwoman who picked me up was ready with her kindness and her delight, pleased to hear I was visiting Sister Irene, with whom she was well acquainted and very fond. She was in raptures that I could speak English so clearly for an American, and perhaps because of my chameleon accent and my civil replies to her questions about how I liked her part of the world, she sounded me out about whether I'd like to be married off to a local farmer. Wives, apparently, were scarce here. She praised the life of a farmer's wife, though she admitted it wasn't easy, and dropped me off at the gate. The sign at the gate said:

4:30am vigil
12:30 dinner
2:30–5 visitors
5: prayer
8: retire

or something very close. I had plenty of time.

Alice told me her great-grandfather – her mother's mother's father – came over as a small child, late last century, one of a dozen or so children brought over by her great-great-grandparents, and it was her grandmother, the child of this infant emigrant, who would tell them O'Malley stories. O'Malley was her grandmother's maiden name, and she had taken it up as an adult, in continuing identification with the heroic figure who loomed over her childhood. Being descended from this undomesticated heroine impressed her deeply, though other relatives – her grandmother's brother, for example – disavowed the connection, and her mother tried to talk her out of it. She was six feet tall and had a beard, her mother would say. Is that what you want? Or, She killed her husbands. You like that? Both of us did, though Grace outlived both her husbands without recourse to murder.

Born in the 1530s, she lived until 1603, to the end of the independent Irish aristocracy, raided ships, had sons, lost fortunes, dealt with Queen Elizabeth as an equal, and left a legacy of legends, some of them true. Through intelligence and force of will, she had seized more power than women ordinarily possessed in her part of Europe, had a fleet of trading and raiding boats that went to Spain to exchange Ireland's fish, hides, and fabrics for Spanish wine and salt and had forced boats on the way to Galway port to pay her tribute, and her adventures on land were lively too. It is a wonder she survived seventy years of wars, raids, conquests, captures, and shifts in political power. In 1838 a writer conducting an ordnance survey of County Mayo wrote, "She is now most vividly remembered by tradition, and people were living in the last generation who conversed with people that knew her personally. Charles Cormick of Erris, now 74 years and 6 weeks old, saw and conversed with Elizabeth O'Donnell of Newtown within the Mullet, who died about 65 years ago who had seen and intimately known a Mr. Walsh who remembered Grainne [one of the Irish versions of Grace O'Malley's name]. Walsh died at the age of 107, and his father was the same age as Grainne."

That Alice's family had emigrated from Clare half a century or so later with the memory of Grace intact, and that it had survived down four American generations, impressed me as much as the earlier generations who remembered her at only third hand more than two centuries later. This child of an emigration as long ago and as many generations away as my own family's was still steeped in identification with a country to which she'd never been, but then Alice had been raised as Irish and Catholic and eastern, all the things that had come out in my wash of outmarriage and westward migration. I was only what the Irish might once have been and are no longer, a wandering barbarian from the western fringes of the world.

I told her about Roger Casement and she told me about Grace O'Malley over our sodden breakfasts. Alice had been told that Grace cropped her hair when young and stowed away on one of her father's boats bound on a long voyage, only revealing herself when it was too late to turn back. Legend has it that she became a sailor, and she did, in the historical record, remain a woman of the seas all her life, commander of fleets and nautical battles (in the story I had read, she gave birth to her second son on a ship and came aboveboard shortly afterwards to do battle). Another was that she took a liking to an aristocratic boy and kidnapped him, perhaps an evolution from a couple of more historically documented stories. In one of them, she rescued a young man from a boat shipwrecked near Clare Island and fell in love with him – but the MacMahons killed him while he was deerhunting, and she slew the responsible parties in revenge. In the other, she was refused hospitality by the lord of Howth Castle, near Dublin, so she kidnapped the young heir, but surrendered him upon the promise that the table at Howth would always have an extra place set. It still does, as Alice's aunt ascertained on a visit.

Grace O'Malley, or Grainaille, has become, with the legendary Queen Maeve, Countess Markievicz, and President Mary Robinson, one of the icons of strong women in Irish culture, and she is talismanic with many who aren't O'Malleys. Although the earlier Celts had made a little more room for women than have most societies, by Grace's time the room had narrowed, and Grace was exceptional. Her power was what she took, not what she received. There was a story about Grace and a hermit too, that her biographer, Anne Chambers, tells, a story that took place on an island in Clew Bay. "A chieftain of a neighboring clan, who had been defeated by Grace, took refuge in the church on the tiny island which was inhabited by a holy hermit. Grace, in her determination to prevent the chieftain's escape, surrounded the church and waited to starve him out. The chieftain, however, with the aid of the hermit, dug a tunnel out to the steep cliff-face, considered impassable, but by the aid of a rope managed to lower himself down the sheer rock-face to a waiting boat. . . . The hermit, breaking his vow of silence, came out later and informed the waiting Grace that her quarry had escaped and admonished her. . . ."

The hermitess lived by a stream, in an indentation in one of those terrible pine plantations owned by the government, not far from the road. Her ducks, her goat, and its kid wandered around the grounds, and the whole place looked more like a bustling farm and less like a remote fastness than I'd expected. Two little boxlike houses, one red, one gray, sat on the slope above, and a cottage with a cross on it sat down below. More or less a hut in a clearing, but St Anthony was never tempted in a landscape like this. The hermitess herself was theatrically good-looking, like Jean Seberg in *Joan of Arc*. She was young, small, had beautiful skin, and brown bangs sticking out from under her wimple of dark blue, a cassock of some heavy grayish stuff with a blue Celtic cross on it, and a turtleneck underneath, to match the thick socks inside her sandals – and the thick gold band of a bride of Christ on her ring finger. She was in the lower cottage, talking to a man about the heating. He was extremely tall, he was wearing a good suit, and, even after I joined them, he kept making forlorn attempts to get us to talk about the World Cup. It was scheduled to begin in a couple of weeks, and most of the country was heating up to a monomaniacal fever about Ireland's chances in it, but the poor man had probably found the two people in Ireland least interested in football.

He did better talking with us about church history and heating. We toured the chapel, which was being built out of cinder blocks. We had tea, or at least we the guests did, for the funny little teapot only held two cups, and she passed around a huge tin box of cookies that must have been a gift. She seemed much more worldly than I expected, and for a recluse was quite alert to the politics of her church. She spoke with asperity about the lack of ecclesiastical support for the contemplative orders, in contrast to those which do good works in the world, and she intimated that contemplation and praying for the world rather than working in it were the real purpose of a religious vocation. There was a great rise, she said, of hermits in western Europe – there were a hundred in Britain and more than that in France, which is undergoing a real revival of the stricter orders – and in Ireland, seven hermits and hermitesses. Perhaps she seemed worldly because I wasn't religious after her fashion and didn't know how to speak to her of the unworldly things, and my interest in her choice of solitude, in the spirituality of solitude and immobility, seemed too intrusive after the talk of furnaces and football. But the tea and the image of hermits and hermitesses proliferating in the out-of-the-way places around Europe sustained me for the rest of my walk back, against the wind.

Alice marched with New York's Irish Lesbian and Gay Organization every year, and when I spoke with her, she seemed to have gotten used to the idea of spending St Patrick's Day in jail. The ILGO was founded in a Japanese restaurant in 1990, and after its members "had lots of fun in the summer's gay and lesbian parade," as another member put it, they thought they'd try the city's huge St Patrick's Day parade too. That's when their trouble began. Cardinal O'Connor used their desire to participate as an occasion to inveigh a little more against homosexuality. The Ancient Order of Hibernians, the parade's Catholic organizers, claimed that the parade was full and they would be put on the waiting list – but when the ILGO sued for their right to march, the Hibernians couldn't produce any such list. A progressive Bronx division of the AOH invited the ILGO to march with their banner in that St Patrick's Day parade in 1991, the city's 230th annual such commemoration. In a show of solidarity, the city's African-American mayor, David Dinkins, marched with the two hundred gay and lesbian participants rather than at the head of the parade as New York's mayors traditionally did. It was unclear whether the mayor or the people he was with elicited the forty blocks of boos.

In 1992 things didn't go so smoothly. In 1993 the event was nearly canceled, Dinkins stayed away in protest, and more than two hundred gay and lesbian protesters were arrested for showing up anyway. In 1994 the veterans who organize Boston's St Patrick's Day parade canceled it after a similar Irish queer group there won the right to participate, and a hundred gay and lesbian activists were arrested while protesting their exclusion. New York's Ancient Order of Hibernians went to court and redefined their event as a religious procession rather than an ethnic parade, to give themselves a legal standing with which to exclude the ILGO or any other group they disagreed with. It was less an Irish than an Irish-American ruckus, the ILGO organizer Susan O'Brien told me in a voice with the soft remnants of an Irish lilt; the celebrations of the saint's day in Dublin and Cork allow gays and lesbians to participate without fuss (and in Cork a lesbian and gay contingent that carried a "Hello, New York!" banner won the "best new entry" award). It was as though the New York and Boston parade organizers were trying to narrow down the definition of Irishness or seeing it in nostalgic soft focus, freed from the complexities and contradictions of Ireland as a real place. Their parades seemed to propose that participants must choose whether to be queer or Irish; to try to be both at the same time meant spending St Patrick's Day in jail, a peculiar commemoration of a saint who had been a slave and a country whose principal heroes had been prisoners and exiles. Jail was, added O'Brien, not particularly unpleasant, what with all your friends there with you.

After the hermitess's home, the landscape on either side of my road became more beautiful – rough country, scrubby and steep, with rivulets and streams. The winding road went up and over a low shoulder of Croagh Patrick, the shrouded mountain that had been on my right all day as I rotated clockwise round it, and when I reached the crest of that shoulder, I could see the bay. Sheep and cattle roamed free across the lonely road, there were minor waterfalls and few buildings until I came close to the bay again where, most of the way round my twenty- or twenty-five-mile route, I stuck out my thumb and got a ride into town from some French travelers.

St Patrick was fasting on the mountain. Grace O'Malley was sailing the seas. Alice O'Malley was marching in a parade that was trying to walk backwards. Sister Irene was seeing to her goats or making tea. I was walking through their worlds. Grace was in the grave, though it's not clear which one. Alice was spending St Patrick's Day in a New York jail cell. The hermitess was on the edge of a tree plantation praying for the rest of us.

16 Travellers

I couldn't tell when my ride was supposed to come to an end. I had been keeping an eye on a young woman with heavy gold hoops in her ears I thought might be a Traveller, but she got off before we came to anything that looked like a prison, and so I asked the bus driver where Wheatfields Prison was. When he found out I was looking for the Clondalkin Traveller site at the walls of the prison his middleaged face hitherto as bland as a sofa cushion bunched up in fury. "Why are you interested in *them*?" he demanded, and I said noncommittal things. His rage increased and he said that they had killed the son of his friend "and the boy was just going on fifteen, he was a lovely lad. And they sat outside the courthouse laughing and drinking. Him as did it got only nine months but the boy's gone forever. As far as I'm concerned, they're the scum of the earth." All of them, I asked, and he said yes. I asked him if he'd ever spoken to any of them, and he said he didn't need to, and why was I going to? I'd been invited to visit, I said, and thought I'd see for myself. Not for nothing was I raised by a fair-housing activist, and I threw in a few platitudes about not judging a whole population by the actions of an individual.

Hate had entered my holiday, along with nomads. My travels up the west coast of Ireland had come to an end with a question about Travellers, a question only Dublin seemed capable of answering, and Dublin had answered it with a swarm of facts, a few encounters, and an invitation to visit a Traveller family. We had come to the end of the line on this bus route from downtown Dublin, and the bus driver insisted that I stand up and hold onto the pole next to his seat so he could give me a private tour of the suburban tragedy he was so bitter about. With an angry sweep of the arm he showed me the bare earth rectangle full of bare new blocks of houses where his friend lived. And as he turned around on his route on this muggy afternoon, he showed me the cemetery where the boy was buried and the camp where the killer – an inadvertent killer by means of drunk driving – lived; I wondered if the driver had done his time in Wheatfields Prison so that the whole story was all at the end of this bus line.

The bus driver also wanted to show me where they – some Travellers, but to him all Travellers – had broken into a waterpipe, "and if you or I did it we'd be punished for it." He seemed as furious that they weren't paying for the water as that his friend's son had died. We passed a wide lawn with a row of little concrete plugs along its perimeter to prevent any Travellers from pulling over to camp and, on the other side of the road, a pipe pointing straight up and trickling water and an encampment of a few trailers with debris scattered around them. If this is what freeloading looked like, it didn't look very luxurious. He finally let me go on a nondescript road with directions to walk down it until I saw the prison – and the admonition, Be careful, it's a mugger's paradise. With its weedy bulldozer heaps and brand new rows of identical houses, it looked more like the road to nowhere, and I set off down it with my offering of peaches and cherries weighing heavier and heavier on this hot June day.

All along my meander up the west coast the people I met had been murmuring stray facts and opinions about Travellers, as Ireland's indigenous nomadic people are currently called. A woman who apparently lived in a trailer herself, outside one of the hostels in Bantry, had told me that they were grand people if you got to know them, though few enough did. While I was walking with the giantess in Ennis a sandyhaired boy of nine or ten had begged change from us and acknowledged he was a Traveller when I gave him a coin, but he was too cringing to tell me more. When the man who drove me to Galway had exhausted the subject of stone walls I'd asked him about Travellers. He told me a story about how Galway is divided into four quadrants, and each quadrant had been shirking responsibility for building a halting site for Travellers for so long that the bishop of Galway had offered the land next to his palace. I had seen Travellers' big black and white carthorses grazing by the dump near Portumna and a long row of trailers on the narrow shoulder of the road back to Galway. Sister Kathleen had said her family always had a load of turf or a can of milk for the Travellers who came by the farm when she was a girl and they were called Tinkers. Bride in Westport had amplified that, saying the primitiveness of the Travellers' lifestyle was only contextual and recent; she herself had grown up in a farmhouse without running water, and they got along well with the Travellers who came along then. Later on, Lee in Ballydehob wrote me a letter about encountering a Travelling man who was standing up in his cart and driving his team of heavy horses at a full gallop, with a gleam of joy and sense of power in his face, and their eyes met in a moment of camaraderie and recognition. Hated, isolated, and sometimes admired, but why?

Finding out in Dublin wasn't as easy as I thought. Although everyone in Ireland and Great Britain seems to know about Travellers – know at least more than I did, and as much as they thought necessary – no one seems to think the subject is interesting, and little has been written on them. They seem neither exotic enough to garner much anthropological attention nor homogeneous enough to be included in national folklore researches; and virtually no books had been written about them before the 1970s, and few enough after that. My first foray in Dublin was instead an education in how the nonnomadic community responds to them. None of the major libraries seemed to have useful books, and the Irish newspapers were not indexed or microfilmed. After I found the one book available on Travellers in the biggest bookstore downtown – a vivid oral history by a Traveller woman, Nan Joyce, who's become a Travellers' rights advocate – I talked my way into the archives of *The Irish Times.*

The Irish Times is the weightiest, most official-seeming national newspaper, like *The Times* of London and *The New York Times.* The head archivist was very obliging on the telephone and invited me to come up to the clipping room. He sighed when I got there, and told me how behind the times Ireland was technologically – You wouldn't believe how recently the paper stopped using hot type he said, then showed me how they kept track of stories. A few people with big shears sat at cluttered desks, dissecting each day's newspaper and pasting the sorted-out stories into a whole library of colossal scrapbooks with their subjects handwritten along the spine, and bits of newspaper lay everywhere like autumn leaves. I sat down to read the last year in the public lives of the Irish Travellers, and as stiff page after page of clippings went by like entries in a national diary, a picture of a civil rights war formed.

On 13 July 1993, a shop on Grafton Street had refused to sell ice cream to a Traveller boy. On 13 October, fourteen windows were smashed and two vans overturned at Four Roads Pub in Glenmaddy, County Galway, by a crowd of more than a hundred people angry about the presence of Travellers. On 23 October, *The Irish Times* declared there were, according to 1992 figures, about 23,000 Travellers in Ireland in 3,828 families, and more than 1,100 lived on the roadside – that is, as nomads. "There are also believed to be about 15,000 Irish-born Travellers living in Britain and 10,000 Travellers of Irish descent in the U.S." On 6 November, the news was that "Clubs, public houses, and shops will be barred from discriminating against Travellers under legislation being prepared by the Department of Equality and Law Reform," which told me in a round-about way that such discrimination existed and was legal.

On 15 November, "The dumping of dozens of mounds of a foul-smelling fertilizer beside Travellers' caravans on the Fonthill road in North Clondalkin last week has aroused anger among the Travellers there, who claim the dumping is a thinly disguised attempt by Dublin County Council to force them to move." This, I would find, was a common story; roadside Ireland's landscape was being redesigned with barriers of stone and earth to eliminate the Travellers who hadn't been eliminated by regulations; a whole zone between those in motion and those in homes was being eliminated. On 18 January 1994, *The Irish Times* noted, "Tension, and sometimes open conflict, between the travelling and settled communities have been long-standing blemishes on society in this State." The following day provided an example: "Publican Threatened with Loss of Her License for Serving Travellers," said the headline. On 8 February, the bishop of Galway had donated land beside his residence for six families' halting site. On 28 February, the Irish Traveller Movement itself became embroiled in controversy "over one of its key policies, that Travellers should be regarded as a distinct ethnic group."

But the biggest and nastiest story in *The Irish Times* was the most recent. In Navan, County Meath, I gathered from the spotty coverage, a group of twenty-six Traveller families had camped next to a school and attempted to enroll their children in it. On 27 April, the paper reported, "Mrs Nell McDonagh sat in her car yesterday morning, crying. She came to remove her child Steven (15). '. . . on the day people are getting the vote in South Africa, why should we have to take our children out of a school? You cannot force your child into a situation where they are not wanted.'" The following day the news was that "about 350 students not taking exams have been withdrawn from the school in protest at the continuing presence of the 26 Travelling families outside the school gates," apparently in a bid to shut the school down or at least keep it segregated. On 2 May, the news got rougher. "Some 40 local residents protested at the Travellers' move into the town's one official site on the Athboy Road, and while there was no violence, gardai [Ireland's police] said the protest was 'pretty tense' at times. . . . The chairman of the Combined Residents Association, Mr. Andrew Brennan, said the situation with the Travellers was a 'powder keg,' adding that the Travellers were claiming to be law-abiding citizens, overlooking the torment and harassment they had forced on people living near them. At Mass yesterday, priests around Navan called for 'restraint, compassion and tolerance' within the community and condemned the petrol bomb attack in Mr. Stokes' caravan. A number of people left the churches in protest." They were minor stories, not front-page news or features or exposés, just chronicles of everyday conflict, and in my later readings I found that such incidents had been common since at least the sixties. In the early seventies four hundred people had marched to keep a Mrs Furey and her three children out of the Shantalla suburb of Galway, while locals had burned down a house in Moate, County Westmeath, to keep another Travelling family from moving in.

All this news did something to the Irish charm and hospitality I'd met with and all the Irish charity evident in the public concern and relief efforts for exotic crises – didn't undo it but complicated it, like a seam of glittering quartz running through soft granite. To me it was puzzling that the conflict was so widely regarded as insignificant, because it recalled the American civil rights issues of the 1950s and 1960s, when the relationship between races became a national test of values and identity. I wondered whether the Irish have been so used to being history's victims that they can't imagine themselves as the victors, whether the conflict in the North has been so emblematic for many Irish and Irish Americans because it allows them to continue to imagine themselves as the persecuted minority seventy years after Irish Catholics became the ruling majority in the other twenty-six counties. Within those counties it's the realm of sexual morality – a sad parade in recent years of priests' mistresses and molestees, farmgirls strangling illegitimate babies, incest and arguments about divorce and abortion – that brings on national soul-searching, as though the Republic were a postpolitical realm in which only private life remained for the public conscience to address. But the conflict over Travellers' rights is about public space and institutions.

All over Europe, similar versions of the conflict between nomads and the sedentary majority are taking place, and though nomad sympathizers and supporters exist, they are themselves often a minority. Nomads are literally unsettling for sedentary populations, or at least those intent on ethnic nationalism.They move through the continuous landscape of roads rather than within the closed loop of borders, stitching the distances together with their circuits. If nomads are indigenous they disturb the idea of a homogeneous folk with roots in the native soil; if they're not, they're considered invaders – and in many places, several centuries of residence haven't qualified Gypsies as natives in the eyes of their neighbors and sometimes their governments. "Their very existence constituted dissidence," Jean-Pierre Liégeois says of Gypsies, and death, imprisonment, expulsion, enslavement, and forced settlement are among the punishments that have been meted out for nomadism in Europe.

From *The Irish Times* I also garnered the addresses of the principal Travellers' rights organizations in the country. There I finally began to see something of the culture that was eliciting this upheaval. At the Dublin Travellers Education and Development Group, on a seedy square outside the center of town, the organizers showed me slides and sold me books and introduced me to the shy Traveller girl who worked downstairs with her brother, gathering and redistributing scraps and remnants the schools used as art supplies. She replied reluctantly to my questions and then suddenly proffered the statement that Traveller society was like Moslem society in its constraints upon women. At the Parish of the Travelling People on Cook Street – a parish without borders, founded because the geographically defined parishes didn't serve Travellers' needs – back near the Liffey and the center of town, they ushered me into their library and let me photocopy away.

A garrulous priest's assistant kept interrupting my reading with stories of his own about working with Travellers. They were, he told me, devout, but magical in their beliefs: more interested in the sacraments and miracles of the church than in the morality, and they often sent their children to school only long enough to receive the instructions necessary for first communion (which is provided by the schools, in this nation of linked church and state). He had seen schools in which the Traveller children were sequestered in rooms with heavy curtains and given different schedules from the rest, and less rigorous classes, lest other parents withdraw their children as they had in Navan. And he told me more, about the tradition of first-cousin marriage and of marrying very young – fifteen or sixteen for the girls – and of the way the parish was trying to discourage these customs. He spoke of Travellers fondly, as though they were children, good but in need of guidance. And the efficient women behind the desks in the other room sent in Cathleen McDonagh, a Travelling woman of about my age, from whom I learned the most, the one who invited me to Clondalkin in the Dublin suburbs.

The background that came to me piecemeal, through all my hunting in Dublin and long afterward at home, looks something like this: no one knows exactly at what point Travellers emerged from the rest of Irish society. The term *Traveller* itself has been accepted in the last few decades as the Travellers' own more civil alternative to *Tinker*, a word which like *Negro* has become derogatory, and to *itinerant*, with its social-worker overtones. Too, Travelling, as it is sometimes capitalized, is foundational to the group's distinct identity, unlike the fading craft of tinsmithing or tinkering. Those who consider nomadism a deviant or dissolute way of life often suggest that Travellers are nothing more than refugees from the economic crises of the potato Famine and perhaps of Cromwell, people who took to the road as beggars and don't know how to get off the road. The idea that it is a very recent way of life or not a way of life, a culture, at all, that it is only a crisis condition of marginal and subnormal people, accords well with the idea that Travelling is a problem to which integration into sedentary life is the solution. A group of Travellers emigrated to the United States during the Famine and remains a distinct group in Georgia, retaining some of the nomadism, language, and other ethnic hallmarks, making it clear that the culture or ethnicity was fully developed a century and a half ago.

Travellers themselves sometimes tell a story akin to that of the Wandering

Jew, in which they are the descendants of the metalworker who made the nails for the Crucifixion and for that deed were sentenced to wander the earth until the end of time. In her study of Travellers, Artelia Court proposes possible links to the outcasts and wandering craftspeople of pre-Christian Celtic society in Ireland. External evidence suggests that some version of the Travellers existed as far back as the twelfth century, when references to "tinklers" and "tynkers" appear; an English law against "wandering Irish" was passed in 1243. Travellers have a language or dialect of their own called *cant, shelta,* or *gammon,* which scrambles words of Irish and English derivation, and one of the strongest arguments for the ancientness of the culture is that their word for priest, *cuinne,* is an old term for druids, otherwise known only from ancient manuscripts. Other linguistic elements suggest roots before the twelfth century. But all the evidence is slight: there are clearly wandering craftspeople and beggars and references that mingle Gypsies and Tinkers from the sixteenth century onward, but there are few details. Tinkers are not Gypsies; they are as fair-skinned and Catholic as anyone in Ireland, and it has been proposed that though Gypsies spread all the way from their origins in India to England, they never reached Ireland because their commercial-nomad niche was already filled by Tinkers. At the turn of the twentieth century, Synge wrote of Tinkers along with all other denizens of the road, but the distinctions are blurred. It seems as though so many groups were wandering the roads for so many reasons that Travellers didn't stand out very dramatically, until everyone else stopped moving.

It is now a matter of debate whether they constitute a distinct ethnic group. Some Travellers seem to want the legal protection and cultural recognition such an identity would confer; others, to think that such status would further alienate them from the mainstream of Irish life. In their report on the ethnic issue, the National Federation of the Irish Travelling People declared, "In their deep religious feeling, generosity and attachment to the family, Travellers have clung to aspects of Irish life to a far greater extent than the settled community." Much of what the sedentary Irish say about the Travellers is what the English and Anglo-Americans once said about the Irish: they drink, they brawl, they have too many children and too little work ethic, they're improvident, dirty, and lawless. The very terms in which the sedentary speak suggest that the Travellers have preserved the tribal and not yet European culture of an earlier Ireland.

Kerby Miller, in his history of Irish emigration to North America, writes about the ways in which the Catholic Irish were at odds with the industrial and Protestant-dominated societies they found themselves in: they "seemed so premodern that to bourgeois observers from business-minded cultures, the native Irish often appeared 'feckless,' 'childlike' and 'irresponsible'. . . . The shrewdest recognized that ancient communal values and work habits persisted despite commercialization. . . . In addition, the Catholic lower classes seemed to lack bourgeois concepts of time and deferred gratification. . . ." Court writes of the "antiquated traditions and artifacts that had vanished elsewhere but which Ireland possessed in abundance" after the Second World War, adding, "And even among these countrymen the Tinkers were conspicuous for remaining doggedly true to themselves." "We are Irish," insisted placards at some Travellers' rights demonstrations in the 1980s, since their differences from the mainstream were regarded as alien rather than anachronistic. It may be that the Travellers stand out for not having changed enough in a society that has transformed itself radically in the last several decades.

Though possessing ancient origins may, for the sedentary scholars of Travellers, confer greater legitimacy on them, it is apparently of less interest to the subjects themselves. The authenticity of origins, the historical basis for identity, may not be their method. That notion more than almost anything convinced me that they did constitute a distinct culture or subculture in this history-haunted place. The anthropologist Sinéad Ní Shuinear writes, "Some nomadic peoples – the Jews of the Old Testament spring immediately to mind – cultivate both literacy and historical memory. Others, even without literacy, enshrine genealogy and significant events into formal litanies to be memorised and passed on verbatim by specialists. But others still – and this includes most commercial nomadic groups – treat the past itself as a sort of baggage which would tie them down in the present. Instead, they cultivate an intense present-time orientation, living in a perpetual now, deriving their sense of identity not from taproots deep into the past, but from vast networks of living kin. The essence of Gypsy and Traveller culture is its fluidity. Gypsies and Travellers everywhere are supremely indifferent to their own origins." She cites the Italian anthropologist, Leonardo Piasere, who "argues that Gypsies and Travellers are not ignorant illiterates, but have very deliberately rejected literacy, knowing that it would solidify the past, thus imposing a baggage of precedent curtailing flexibility in the present." Like the Western Shoshone and other nomads, Travellers traditionally destroy all the belongings of a person who has died, a process that tends to rule out heirlooms and vast accumulation, a means of keeping its practitioners even materially in the present.

The French theorists and nomad enthusiasts Deleuze and Guattari declare that the hierarchical model of the tree has dominated too much of Western thought and offer in its place the rhizome, the loosely structured, horizontally spreading root system of plants such as strawberries; Ní Shuinear's proposal that Travellers are organized socially and imaginatively around contemporary networks rather than historical taproots echoes their metaphor.

The intimation of such a radically divergent sense of time, space, and society electrified me, but other information and conversation tempered my romanticism. The enormous contemporary enthusiasm for nomads – the romanticism that has brought into being so many boutiques, tattoo parlors, artists' projects, pseudoethnic recordings, and books with "nomad" in their names – is premised on the dubious idea that nomads embody on a mass scale the freedom of the solitary traveler, that romantic figure silhouetted against an exotic landscape like the individualist tree. For those of us who are largely sedentary, travel is a way out of the world that surrounds us, but nomads rarely if ever leave their world: it moves with them. The Traveller activist Michael McDonagh explains that "for Travellers, the physical fact of moving is just one aspect of a nomadic mind-set that permeates every aspect of our lives. Nomadism entails a way of looking at the world, a different way of perceiving things, a different attitude to accommodation, to work, and life in general. Just as settled people remain settled people even when they travel, Travellers remain Travellers even when they are not travelling."

The spatial freedom that might otherwise dissolve their society and identity altogether as it does for us temporarily, as respite, vacation, and escape, is counterbalanced by a greater rigidity of social structure. Architecture and geography hold our lives in place – identities built into the layout of the house, the status of the address, and the routine of the day – but custom alone must hold theirs in lieu of place and therefore must surround them as surely and solidly as a locale. The nomad's fluidity of time and space and work and property all occur within a stubbornly conservative culture perpetuated in tightly knit families. Likewise extreme feats of travel have little to do with nomads. It is exhilarating that individuals should walk the length of a continent or carry a sixty-pound pack over the remote mountains, but such feats are for solitary adventurers in their prime, not for groups for whom travel is a permanent condition including all the goods and generations, and certainly not for commercial nomads like the Travellers and Gypsies who earn their living from interactions with the sedentary community. Still, one romantic attribute remains, that of movement itself, of the constantly shifting scene, the unpredictable life lived closer to the bone of those in motion, uninsulated by the buildings and goods and familiarity of settled life: that is a romanticism Travellers and sedentary people seem to share.

Travellers have traditionally been self-employed or temporarily employed, surviving on a plethora of skills and talents, and often shifting roles and images to accord with the work – abjection for begging, an air of responsibility for contract labor. Indeed much of what seems to be considered Traveller and Gypsy dishonesty is the art of saying what works or pleases in wildly varying and often hostile circumstances. Taking permanent jobs conflicts with the fluid autonomy of their identity, argue some of the sociologists of Travelling. They may be the last people in the industrialized world to have collectively escaped wage labor, escaped selling their time and setting their lives to someone else's schedule, but the price of their social redemption seems to be the surrender of the fluidity of their labor, spatial, and temporal structures via the taking of jobs. By many accounts and oral histories, Travellers who get ahead often take it as an opportunity to take time off or travel, disregarding the longterm security to which the wage earner aspires. Freya Stark, the travel writer who spent years among the pastoralist nomads of the Muslim world, writes, "The life of insecurity is the nomad's achievement. He does not try, like our building world, to believe in a stability which is non-existent; and in his constant movement with the seasons, in the lightness of his hold, puts something right, about which we are constantly wrong. His is in fact the reality, to which the most solid of our structures are illusion; and the ramshackle tents in their crooked gaiety, with cooking pots propped up before them and animals about, show what a current flows round all the stone erections of the ages."

In the picture most accounts paint, Travellers throughout the first half of the twentieth century continued their professions of tinsmithing (from which the term *tinker* comes – the tinkers or tinsmiths made many of the milking cans, buckets, pots, and pans farm families used), horsetraining and trading, begging, fortunetelling, selling balladsheets, handicrafts, and other small items, and working as migrant agricultural and manual labor, encountering hostility and some brutality but at levels that allowed them to continue to be nomadic. They Travelled mostly on backroads and consorted mostly with rural people – one Traveller term for the sedentary is *country people* – though they found work in English cities and the outskirts of Irish towns from time to time. They were appreciated for their skills, wares, and the news and novelty they brought to isolated communities. They were disliked for begging, for sometimes sneaking their horses into farmers' fields and crops to graze, for the dishonesty with which nomads often deal with sedentary people, for theft and suspected theft, and maybe for being an unfamiliar intrusion into familiar landscapes. It isn't clear when they began using barreltop wagons, but they seem to have pitched roadside tents made of hazel branches and some kind of tarp beforehand. Anyone who has encountered the wet Irish land and sky can appreciate how strong the nomadic impulse must be to survive in those circumstances (partially settled Travellers now often say they yearn to roam in the summertime, when the weather is fine).

In the 1950s and 1960s horsedrawn wagons began to be replaced by cars pulling trailers, and horses are now kept more for pleasure than for use (though tourists can rent facsimiles of the wagons, complete with horse, and play Gypsy on the backroads; I had seen such a rental site in Westport). It may have been cars and the concomitant shrinkage of distance and access to manufactured goods that doomed their symbiosis with the countryside; it is often said that plastic did in their way of life. Handmade tinware was replaced by mass-manufactured goods, and as distances shrank and cars replaced horses Travellers' functions as pedlars and horsetraders also eroded. Roadsides were relandscaped to make roadside camping difficult or impossible across the country, unregulated space dried up, and they began to stay longer wherever they were, however unwelcome, since finding another halting site would be hard. Ironically, hostility seems to make them stay rather than go. There are regulations that halting sites and housing must be built for them, but many projects have been delayed by opposition and many that do exist are inadequate or inappropriate in their design. In the 1970s a Traveller told an oral historian, "For a woman a house is a grand thing for her to put the children in. But for a man a house is only a payment of rents. . . . Lots of travellers have houses in the winter and they leaves them lonely when they take to the roads in the summer. It's only a bother having the house and it's not healthy to be shut inside them four walls with no trees in sight and only the windys to keep you half breathing. No, I'd sleep in a stables before I'd sleep in a house. . . ."

Travellers now appear to be something of a displaced population, in flight from the destruction of rural life as much as any farm people, but they are seen less as refugees than intruders wherever they go. Many have ended up on the periphery of towns and cities, and some have gone to England. The programs of forcing them to settle into fixed houses are over, but the elimination of the necessary sites and circumstances for Travelling continues, as do some voluntary housing programs. Though Travellers are central to the scrap metal and car parts industries in Ireland, and a few Travelling families have become wealthy antique dealers (and, because of their wealth, are seldom counted as Travellers), a high percentage are on the dole. In a country with more than 20 percent unemployment overall and intense exclusion of Travellers from all institutions, welfare dependency is not surprising. But in addition to the suspicion nomads and minorities usually attract, the Travellers are now hated with the peculiar fury taxpayers reserve for those they consider freeloaders.

Most recently the authorities have become better at navigating a middle ground, of providing housing adapted to the needs and customs of these increasingly immobilized nomads. In such a housing complex did Cathleen McDonagh and her family live, up against the walls of Wheatfield Prison. I recognized it from the Travellers' parish worker's description: a double row of diminutive houses with wide driveways, arranged in two lines flanking a central green, with a high gray prison wall behind looking like the back of a stage set. When I reached the grass a group of little boys ran up to greet me and inspect me. They were tough, scruffy, but polite, and I could tell I wouldn't get far without their cooperation. So I told the one who seemed to be the leader, a stout, chestnut-haired boy of about ten in an undershirt, who I was looking for. She's my cousin, he said, and began to lead me to her trailer. The boys asked if I was a social worker, and I told them that I was a writer from America. I knew that would keep them busy for a while and it did; they too had to know who I'd back in the World Cup. A middleaged man came up to us, another inspector; I introduced myself and we shook hands. It was John McDonagh, Cathleen's father, a powerful-looking, big-bellied man whose mild face gave him a horse's air of harmless power. Cathleen, he told me, was in her sister's house, and so we doubled back the way the boys had led me, and he took me into a kitchen with a plump woman – the sister – washing lettuce, a child in a high chair and another one roaming around the tiny room, and Cathleen sitting and talking. She showed me around the tidy house, which was bigger than it looked from outside, showed me her nieces sitting on the edge of their bed knitting and looking very diligent for girls of twelve or so, and took me into the parlor. I perched among the fat lace pillows of the sofa, facing the corner cabinet of richly colored dishes, the mantelpiece's two plates depicting a horse fair, and my hostess.

In her cut-off jeans and black t-shirt, she looked much much more at ease than in the long patterned skirt in the Dublin office. I had met her at the Travellers' Parish, where she was studying to advance her education beyond the primary-school level where she had left off, and to gain the skills to become a Travellers' rights advocate. She was a bigboned, broadshouldered woman of my own age – early thirties – with high cheekbones and powerful pale blue eyes beneath her thick brown hair. In the parlor, amid the lace and china, she continued to talk of prejudice in the low, flat voice she'd used before, a voice that sounded both cowed and resistant. She spoke in examples rather than abstractions. She spoke of how every Traveller is held accountable for the acts of any one of them. Of how when Travellers misbehave, they tend to do so in public – almost every aspect of their lives is much more visible, outdoors and by the roadside – and thus gain an exaggerated reputation for drinking and brawling. Of how they don't want special treatment, only the rights of the rest of the citizenry: access to the same education, entry to the same places, housing or at least halting sites. She told me about last

Christmas when her brother came over from England. It's customary, she said, to do a good deal of celebrating around Christmas, and so she spent all she had on a disco outfit. But when they got to the club in Dundalk, they were told they would have to wait because there were too many already inside. But there were people all around them pouring in. Apartheid Irish style. You get very guarded after experiences like that, which is why Travellers might not be easy to get to know. People think they're rich because of their vans and jewelry, but they buy the vans on credit and need them for their work, just as settled people do their houses, and the jewelry is akin to savings. All her own jewelry, she told me, was gifts – the three gold bangles from her parents, the big gold hoop earrings from her brother in England. She frequently ended her sentences Please God, to indicate that her desire or ambition was tempered by God's approval, and her religion was an important part of her life.

The ice broken or at least a little thawed, we went into her trailer – her parents, she said, had allowed her to remain unmarried, and she had a trailer of her own – and she began to have real conversation with me in a different, more natural voice. Her brother and other men kept dropping by to say hello and inspect me, and I met her younger brother William and inspected the dagger tattooed on his forearm in return. The trailer, the kind that hitches to the back of a car, was the size of a small room, and everything in it was neatly arranged, the bed folded back into a couch. The clock she had broken that morning, however, was lying on a counter all in pieces. She had beautiful dishes arrayed on narrow shelves above the windows and two books on another counter: Bruce Chatwin's *The Songlines* and Peter Matthiessen's *Indian Country*.

It turned out she was interested in Native Americans and identified with them, with better grounds than most who do. She had acutely picked out their nomadism as one of the reasons why Euro-Americans considered them barbaric and one of the grounds for persecuting them. Civilizing Indians usually entailed transforming them from nomads into agriculturalists, tied to the success or failure of individual labor on a small piece of land. The sedentary toil of agriculture was usually considered inseparable from or foundational for culture itself by the nineteenth-century Americans who made Indian policy; what they would think of the present's postagricultural societies is hard to imagine. Turning Native American nomads into agriculturalists largely failed, but it did succeed in justifying a drastically reduced land base for them and thereby freeing up much of their land for others. My hostess also pointed out that what the Jews and Gypsies killed in the Nazi holocaust had in common was nomadism, and I was glad I'd introduced myself as a Jew and might thereby be considered an honorary nomad. (I had somewhere around Galway stopped telling people of my mixed ancestry, because it was clear I wasn't Irish in the way the Irish were, because trying to explain that being mixed didn't mean being nothing was getting tiresome, and because declaring Jewishness dried up any further questions. I thought I'd probably only really ever feel Irish Catholic if I went to Israel.)

Each door Cathleen led me through seemed to take me into a more personal sphere; we had moved from the formal interview on her sister's sofa to the conversation in her trailer to the festive visit in her parents' trailer next door. I seemed to have been accepted, at least as a guest. In her parents' trailer Cathleen laid out thinly cut fresh bread, cold meats, and tomatoes and began to make cup after cup of strong tea for us all, washing the cups thoroughly between each round. Her parents' trailer was airy and comfortable, a spotless salon of windows, couches, kitchenette, and a central table. What do you call them, they asked me, and I said, Trailers. They looked satisfied and said that Travellers too called them trailers; only country people called them caravans. And they asked me about American rest stops; they had heard wondrous stories that the US government built them copiously along the highways and anyone was allowed to halt at them unharassed.

My own country took on new enchantment for me as I told them of the western American infrastructure of rest stops and camp grounds and trailer parks and interstate highways. Of the quite respectable middle-class retirees who sold their houses and took to the road in trailers, migrating like birds alone and in flocks, south in the winter and anywhere in the summer. Of how much of the populace was, if not nomadic, at least restless and rootless, moving on an average of once every five and a half years. Of states where the majority of homes seemed to be prefab trailers that could be trucked to the next location. Of how many fine gradations there are between the absolutely fixed and the fluid in the US, rather than Ireland's stark gap. Of my own adventures in my pickup truck with the shell on the back, traveling around the West, living out of the truck for weeks on end sometimes, traveling sometimes with my younger brother in his pickup when we went to political actions together. As I spoke of days of driving five hundred miles or so alone, of driving a hundred miles down Nevada's secondary highways without seeing another soul, I became homesick for my own roadscapes. Any doubts I'd had about disconnectedness, rootlessness, and fossil fuel economies were bowled over by our collective evocation of the lure of the open road.

I swapped my tales with road stories of theirs, mostly of Mrs McDonagh's. Mrs McDonagh, Cathleen's mother, impressed me as a remarkable woman. Stout and weathered, with her shapeless dress and her graybrown hair pulled back casually, she had made no efforts at beautification but she radiated a calm joy in her expressions and her sweetvoiced stories. Though one might expect a nomad to flicker like a flame, she gave instead the impression of enormous earthy solidity and complete participation in the present. Life seemed to delight her. She told me they could see the mountains from where they were, the Dublin mountains. That her mother was from County Meath, and there was always a town you'd go back to. That home was where your people were buried. And one of the great pleasures of travel was going back to a place in which a significant passage of your life had occurred, revisiting experiences inextricably linked to a distinct locale (unlike, she implied, the sedentary, whose different dramas may all occur on the same thereby unevocative home front). She grew up in the wattle tents – the tents made of tarps and hazel wands – and there would be a big tent with a fire to cook on and sing around. There was a wagon to sleep in (though she didn't make it clear if it was always there along with the wattle tents). She hadn't learned to read or write, which was inconvenient, because you had to ask people to help with your letters, so they always knew your business. When a white moth came and fluttered between us on the couch where we sat, she said, A little moth. That means a letter's coming.

Now it was ten years since they had Travelled, she said, but they went off all the time in vans. And she spoke of their journeys. She had wanted to emigrate to Australia once in the 1960s when visas and jobs were easy to come by, but at the last minute her husband had backed out. He was a less enthusiastic adventurer. She wanted someday to see Russia and Germany, and they had gone on pilgrimages to Knock in Ireland and Lourdes in France. She deplored the long hours of waiting and the poor organization at Lourdes, but they had gone there all along the backroads of France, and the French people they met had been so friendly – a report which was itself testimony to their talent for travel. They were very devout; Cathleen had told me of a pilgrimage to Saintes-Maries-de-la-Mer near Arles in southern France, a place of great Gypsy pilgrimage as well. When her mother recited the places she'd like to go, Cathleen added, Please God, Jerusalem. "Narrow and wide," concluded my notes, "Muslim. Freedom. Change," and I never could make sense of them. Darkness had fallen while we had been swapping stories, and darkness fell very late that time of year. They sent for William, and he drove me back to central Dublin in his van, fast, as I swayed between him and his sister around the bends in the roads.

17 The Green Room

It was warm, sticky, and 16 June when I got up that last morning, the morning after the visit to the McDonaghs. It was, of course, the anniversary of the day *Ulysses* takes place in Dublin on 16 June 1904, the day Joyce is thought to have chosen to commemorate some eventful date at the beginning of his relationship with Nora Barnacle, not long before they left Ireland for good. Bloomsday, as it's now called, has become a citywide if not a national holiday, a mildly ironic institution for the city Joyce abandoned. Or perhaps not, for *Ulysses* alone established Dublin on the map of literary imagination more effectively than all Dickens's novels did London, or Edith Wharton's, New York. Everywhere, piles of brochures announced Bloomsday bicycle races and guided tours and breakfast readings. It was hard to tell whether Bloomsday was a local's or a tourist's event, amid the cottage industry of Joyce statues, plaques, t-shirts, postcards, and other flotsam that, like Bloomsday itself, seemed to function as a pleasant alternative to reading the books. Perhaps the real irony is that the exile Joyce and his fictional Wandering Jew have become key elements in the makeover of Ireland for consumption by tourists, bait for the swarming of populations.

I like holidays, though. They insist that time is cyclical and patterned rather than an uninterrupted industrial flow of workdays, and insist that we celebrate the coincidence or overlap of times. They call upon us to use the present to think of the past as well as the future, and to regard the past as a live force shaping the present. They punctuate the year's long sentence. I had thought about participating in an event or two, since I had coincided with the right place and time, but I had already gone and taken a look at where 7 Eccles Street, Bloom's home, had been, and at a few other sites highlighted on the *Ulysses* map given out by the tourist office. Besides, a ceremony is a religious invocation of the original event, as communion is of the Last Supper, but an imitation is the thing least like an artistic original on the very grounds of originality, no matter their formal resemblances. It seemed to me that morning that the best homage to a secular narrative about people pursuing their own ends would be to pursue mine and listen for the echoes. After all, *Ulysses* is not a novel about a Jew who, in 1904, takes his half-Spanish wife to all Odysseus's ports of call around the Mediterranean. Joyce himself followed Homer very loosely.

So I bought a copy of *The Irish Times* and went to the Elephant and Castle restaurant in the gentrifying Temple Bar district for an American breakfast. My waiter, who had worked at another Elephant and Castle in Manhattan, talked about New York with me; I once had a memorable breakfast at that outpost of the chain on the last morning of a winter visit there, a morning so cold my drying hair turned into icicles on the way to my rendezvous. Now that I had been invoking interstate highways and sagebrush plains and Manhattan canyons, I missed coffee too, but scrambled eggs and June sunshine put me in a pleasant international limbo. It was a typical day in *The Irish Times*: "U.S. Courts Jail Dublin Man in IRA Arms Case"; "First Atlantic Flight Re-Enacted after 75 Years"; "Cleric Cheers Closure of Boutique for Transvestites"; "Over Four Hundred Homeless in 1993: Hard Core of Dublin Young People Now Rely on B&Bs" – this last a story about putting up homeless children in tourist lodgings for extended periods.

One feature in *The Irish Times* seemed intended to mark Bloomsday, however obliquely: it was an article about Ireland's rapidly shrinking Jewish population – though there had been four thousand Jews in Ireland at the time of World War II, there were now less than a thousand, and preserving the culture and finding fellow Jews to marry was becoming ever more difficult. Emigration was the alternative to giving up on being part of that minority culture. One could make up a modern Leopold Bloom recovering, in the style of our times, his ethnicity and his Judaism, emigrating to London in search of community, or following Stephen Dedalus to Paris, or going to Hungary, where his father was from, or going further, to the Mediterranean, to Israel – though half-assimilated isolation was a state he accepted as quite natural in 1904.

I was, that afternoon, going off to spend a few days in Paris myself and looking forward to it. My associations with Ireland were largely symbolic, but I had walked (and flown), not run, away to Paris at seventeen, when seven thousand miles seemed like a suitable distance to put between myself and my life up to that point, as well as a good place to wallow in the historicism California had not yet coughed up for me. I made my first home there and had been happy, because to be alienated in one's own country, in one's own hometown, among one's kin and peers, was problematic, but nothing could be more natural than to be alienated in a foreign country, and so there I had at last naturalized my estrangement. This may be one of the underappreciated pleasures of travel: of being at last legitimately lost and confused. Looking at Ireland as a homeland had unnerved me with the very sense it gave of a homogeneous, predictable, familiar world, until I'd found a roster of colossal changes and an outsider population. At this end of my wanderings I had had enough smalltown friendliness and wall-to-wall white faces too; it was time to mill around in a city that was no longer the French capital of my youth but a polyglot cosmopolis full of Africans and Asians who had reversed the colonial process and were remaking the city, making it less predictable, less conventional, more porous to other possibilities and more complicated identities – more like home for me.

Which is not to say I wasn't satisfied with my trip. Like a beachcomber emptying out her pockets, I laid out the finds of my travels at breakfast: an elephant skull, a Peruvian butterfly, a desert giantess confessing to me over Chinese food, a busride with a Hawaiian woman who explained to me how like her wet island home Ireland was, a steady supply of North American musical traditions and devotees, a New Zealander who had amused me by punctuating every sentence with dead: *dead wrong, dead tasty, dead beautiful.* This last acquaintance impressed me for having wrapped up all his affairs and discarded all his belongings before he left home, so that nothing called him back or limited the possibilities of his trip; travel itself had become a home unshadowed by the necessity of return, a state without boundaries. As my own time away came to a close, I could hear the sirens of familiar comforts and responsibilities begin to call softly.

These were the souvenirs of Ireland as a global crossroads; a month's sequence of local landscapes were likewise pressed like flowers in my notes and mind, saved along with innumerable arrangements of gray stone, twin fountains of flawless tea and whiskey, and episodes of hospitality from dinner at Paddy's sister's in Cork to my repast in the McDonaghs' trailer the night before. I had found what I hadn't known that I was looking for: in the midst of a country that had seemed from afar to be about fixity and memory and literature, a mobile people who had avoided solidifying their past through literacy and history, a real culture that, along with Joyce's fictional wandering Jew, seemed to render even the Catholic Republic of Ireland open to multiple possibilities – of which departure was one. Which is only to say that I had distilled for personal use a fluid version of the past that licensed a volatile present of wanderlust and mixed blood. The fixed and the fluctuant, remembering and forgetting, the pure and the hybrid, roots and wings would sort themselves out and tangle themselves up again, but I had reached a celebratory state where resolution seemed as unnecessary as it was impossible. If I had to spend the rest of my life traveling back and forth between incomplete states, why, I like to roam. And I like inconclusiveness, like a conversation that will always leave more to be said, rather than the conclusion that comes down like a verdict and leaves silence in its wake.

Dublin was good to me in my final days there. I had been meandering with the satisfaction of the return visitor, for whom navigation is no longer a challenge and for whom the pleasures of the familiar begin to gleam shyly among those of the new. Summer had come, the days were warm, the evening light lingered until after ten at night, the bars seemed fuller than ever, and drinkers spilled onto the street, basking in the rosy twilight. People noisily hawked World Cup souvenirs, tobacco, and fruit on the streetcorners, and one haggard old woman selling by the river impressed me as though she had been a witch or a cannibal, with her black perambulator made to carry babies heaped instead with apples and oranges. In the June evenings the arches of the bridges reflected in the Liffey made perfect ovals, like rows of vast eggs through which birds flew and garbage drifted. I had been wandering further afield, beyond what seemed to mark the bounds of the tourist's city, that city which is a phantasm of dislocation and leisure to the resident, just as the resident's working city remains elusive, obscure to the visitor, however much they overlap. In Dublin, of course, they overlap a lot; the center of the city mixes tourist, administrative, and nationalist sites.

One day, amid my Traveller researches, I tried to follow the Liffey on its short course to the sea and found that east of the Custom House and the train station it disappeared behind walls and the streets angled off into a neighborhood of a wholly different texture than centermost Dublin. In no time at all, I was wandering amid cheap modern apartment buildings whose stoops were occupied by plump, poor young mothers watching children rush up and down in the sunshine and grime. I'd blundered into the inhabitant's unglamorous city, away from the Palladian splendor and postcard racks, into a place poor enough to feel dangerous, outside the literary problems set by the texts I knew best, instead inside the terms, say, of Roddy Doyle's novels.

It was the city's central artery, the Liffey, that held my attention, though, and I found a version of it in stone on the walls of the Custom House. Enormously long, gleamingly white, and heavily adorned with sculpture, all but the riverfacing south façade were surrounded by fenced grounds and security guards. But the gatekeeper, once he heard I wanted to look at the art and we had exchanged a few pleasantries about weather and California, beamed and waved me in and sent me to the side entrance to pick up some literature. I wandered round it for a long time, clutching my literature and craning my neck and looking my part. The Custom House had been built on pilings on marshland on the north side of what was once the city's eastern edge, and in its mix of motifs and its position between land and water, it serves as a model gateway between the local and the global. It was built during the brief Golden Age of 1782–1800, the era of the independent Irish Parliament before unification.

An Englishman whose father had been a French Huguenot refugee, James

Gandon, designed it, and he drew on the Roman Doric style for its architecture and the symbolic language of classical allegory to ornament it. Though that language has foreign origins, it speaks of local things in its hybrid tongue of goddesses and harps and cattle, perfect for a customs house which functions, after all, as a filter to adapt the goods of the world to the benefit and custom of the country it serves. The Dáil Éireann, the Irish Parliament, had ordered that the Custom House be attacked in May 1921, because it was "one of the seats of an alien tyranny" that had not yet been replaced throughout the city, and the Dublin Brigade torched it so effectively the fire smoldered for weeks and the stone cracked as it cooled for months. The damaged edifice was originally slated to be torn down, but it is instead a restoration that now serves independent Ireland, though the world grows ever harder to filter out.

Along the river the Custom House has a gorgeous, horizontal sweep of façade that is, in the nice words of the architectural history I had been handed, "beyond praise." Atop the opposite, northern façade, four half-nude statues representing Europe, Asia, Africa, and America stand like victors, the four corners of the world from when it still had distant edges. Commerce, amoral but effective priestess of bordercrossing and transnationalism, stands supreme atop the great central dome, the goddess who shrank that remoteness. But a tourist manual identified her instead as Hope. Probably Tourism herself, Commerce's hopeful sister with the distracted gaze, should be added to the façade to reflect the changing global and local economy; perhaps even Emigration with her winged shoes and exhausted face belongs here – but those goddesses would lead to others, to a little-match-girl icon of Poverty and on and on to a white marble mob of explanations that would crush the roof. Nationalism survives in tiny harps carved into the columns, on the friezes of the porticoes, and harps elsewhere huge and crowned as the Arms of Ireland, holding the flanking English lion and Scottish unicorn in check. Even the motif of the harp, the same harp that appears on Ireland's coins and stamps and passports, is eclipsed by the particulars of Edward Smith's carvings of the principal Irish rivers on fourteen keystone arches on all four sides of the building.

The rivers are represented as staring heads with deeply carved mouths, and they make the immense building a miniature of Ireland, whose rivers flow outward to the sea. Almost everything that is allegorized as a human figure is allegorized as a woman – Liberty, Justice, Commerce, Britannia, Ireland – but thirteen of these strong heads are men or gods whose beards stream and ripple like water and trickle eels and dolphins. They make magnificent allegories. Identifying goods and symbols crown each river: oaks on the handsome Shannon, wool on the beaming Suir, intertwining swans on the happy face of what is probably Belfast's Lagan, tower fortifications on the pugnacious Foyle, which runs through Derry. The fourteenth, the goddess of the Liffey, has the place of honor on the southern façade facing the actual Liffey. Her braids like tamed waters twine around her longnosed, serene face, and her headdress is an abundance of fruit and flowers.

That Sunday of my last week, an introspective day of rest since everything was closed, I walked to Phoenix Park along the Liffey, westward, landward, away from the sea I hadn't reached. On the way back, trying to recall what I'd dreamt, knowing that the whole underlying flavor of the day came from it whether I could retrieve it or not, I saw a ripple in the water, like a thought itself surfacing. The water was green like a bottle, there was a carpet of green waterweed crawling up the walls of the river, and steps and ladders descending down into it, as though for bipedal amphibians, and the ripple became a fish. Thanks to the fish, the surface of the water that had seemed so opaque revealed a degree of transparency to a depth of perhaps a foot. The fish turned into three big fish in a row, blue above and silvery white below, all moving in the same direction I was, and at the same speed. I proceeded for several minutes parallel with the fish, who were perhaps a dozen feet out from the stone embankment of the river, as though we were sharing a Sunday stroll in an interlude in which water and land, swimming and walking, no longer separated us. They became five fish, then the leader took a sudden turn and they all vanished in his wake. Once I knew the secret of looking through the surface, the stagnant-looking green water began to wake up. A cluster of five, perhaps the same five, swam up from the opaque greenness and then swam out of sight again. Nine surfaced, bluer than the cloudy sky, and whenever they turned, they flashed silver. Larger and larger schools began surfacing, as though the whole river were coming to life, but so were the surface and the afternoon light, and they made the water almost opaque again. The last school swam across the river rather than up it, swam into the stately windows of a Georgian housefront reflected upsidedown into the rippling liquid, swam into the wavering windows and disappeared.

The sea was a duller green on the ferry from Rosslare Harbor to Le Havre and the night was pitch black, so that the sound of waves being splashed and plowed, the pitch and the deep hum and throb of the ferry were the only evidence of location. After inspecting all the perfume sales counters and bars and the labyrinthine structure of the boat itself, I settled down for the night. On my uncomfortable couch of two seats and a window ledge, I dreamt I was home again, in my home for the last dozen years, a light-drenched, airy white set of rooms that fits me like a shell fits a snail and that in my dreams constantly mutates in wondrous ways, its familiar doors and windows opening onto unfamiliar landscapes, its rooms periodically growing and its walls cracking to reveal unknown regions and possibilities. The first episode was unspectacular, a complicated culinary dream about making soup on my old white stove and finding a dead bat beside it. It was the vulnerable flesh-pink of the bat's crumpled wings that left the deepest impression when I woke up among the sleepers all around me on the ferry, and in the next dream a friend and I were making a heraldic emblem of a vulture and a honeybee. In the final, unforgettable one, there was a room off the hallway I had never paid attention to or had shut off and forgotten upon moving in, a green sort of sitting room with even a fireplace, where there is in truth nothing but a wall. In this dream on the water between Ireland and France, I had opened it up again and thought I would, with the help of one of my brothers, move all my books into it and have a workroom at last. I woke rested and happy, and the sense of expanded space hovered round me all the way to the continent and beyond.

Notes

Epigraphs

Salman Rushdie, "Imaginary Homelands," in *Imaginary Homelands* (London: Granta Books/Viking Penguin, 1991), p. 12: "It may be argued that the past is a country from which we have all emigrated, that its loss is part of our common humanity. Which seems to me self-evidently true; but I suggest that the writer who is out-of-country may experience this loss in an intensified form."

Matsuo Basho, *The Narrow Road to the Interior*, Sam Hamill, trans. (Boston: Shambhala Press, 1991), p. 3.

1 The Cave

p. 4: Bill's book is: William Studebaker and Max G. Paresic, *Backtracking: Ancient Art of Southern Idaho* (Pocatello, Idaho: Idaho Natural History Museum, 1993).

2 The Book of Invasions

p. 9: "As everyone does, they partitioned. . ." and "On Monday in the beginning . . ." from *Lebor Gabala Erenn: The Book of the Taking of Ireland, Part IV*, R. A. Stewart MacAlister, ed. and trans. (Dublin: Irish Texts Society, 1941), pp. 15 and 203.

p. 9: I was informed of the discovery of this Roman fort by Ray Ryan of Cambridge University Press.

p. 10: "*Britain*, by thee we fell . . ." in Jonathan Swift, "Verses occasioned by the sudden drying up of St. Patrick's Well near Trinity College, Dublin, in 1726," Herbert Davis, ed., *Swift: Poetical Works* (London: Oxford University Press, 1967), p. 385.

p. 11: "It was at best a *mariage de convenance* . . . " in Redcliffe Salaman, *The History and Social Influence of the Potato* (New York: Cambridge University Press, 1985), p. 273.

p. 11: "Ralegh has backed the maid to a tree . . . " in Seamus Heaney, "Ocean's Love to Ireland," *North* (London: Faber & Faber, 1975), p. 46.

p. 13: Dean McCannell compares tourist and military complexes in a *Headlands Journal 1992* interview (Sausalito, Calif.: Headlands Center for the Arts, 1994).

p. 17: "that these Heaps were laid there . . ." in Carole Fabricant, *Swift's Landscape* (Baltimore, Md., and London: Johns Hopkins University Press, 1992), p. 30.

p. 17: "I reckon no man truly miserable . . . " in David Nokes, *Jonathan Swift: A Hypocrite Reversed (A Critical Biography)* (New York: Oxford University Press, 1985), p. 111; and "I choose to be a freeman" in Fabricant, *Swift's Landscape*, p. 52.

p. 18: Re the Irishness of the Brontës, see Edward Chitham, *The Brontës' Irish Background* (London: Macmillan, 1986).

p. 18: Edward Said on Mansfield Park in *Culture and Imperialism* (New York: Alfred A. Knopf, 1993); and Jean Rhys, *Wide Sargasso Sea* (Harmondsworth, England: Penguin, 1988).

3 Noah's Alphabet

p. 21: Recent scholars have speculated that St Patrick's snake-charming . . . in Gerald of Wales (Giraldus Cambrensis), *The History and Topography of Ireland*, John O'Meara, ed. and trans. (Harmondsworth, England: Penguin, 1982), p. 130, note 13.

p. 21: Gogarty . . . long ago loosed some snakes . . . in Hugh Kenner, *A Colder Eye: The Modern Irish Writers* (Harmondsworth, England: Penguin, 1983), p. 252.

p. 24: Elephants, for example, signify . . . in *The Bestiary: A Book of Beasts, Being a Translation from the Latin Book of the Twelfth Century*, T. H. White, trans. (New York: G. B. Putnam's Sons, 1954).

p. 26: "The first subject matter for painting . . ." in "Why Look at Animals," John Berger, *About Looking* (New York: Vintage Books, 1991), pp. 7 and 9.

4 The Butterfly Collector

p. 28: "These people were regarding steadfastly in the direction . . ." James Stephens in "The Insurrection in Dublin" in *Dublin: A Travellers' Compendium*, Thomas and Valerie Pakenham, eds (New York: Atheneum, 1988), pp. 276–8.

p. 29: his father would swim out to sea . . . in B. L. Reid, *The Lives of Roger Casement* (New Haven, Conn: Yale University Press, 1976), p. 4.

p. 30: "I was taught nothing about Ireland . . ." in Peter Singleton-Gates and Maurice Girodias, *The Black Diaries: An Account of Roger Casement's Life and Times with a Collection of His Diaries and Public Writings* (New York: Grove Press, 1959), p. 42.

p. 30: "I have seen him start off into an unspeakable wilderness . . . " Joseph Conrad quoted in Paul Hyland, *The Black Heart: A Voyage into Central Africa* (New York: Henry Holt & Co., 1989), pp. 74–5; and Reid, *The Lives of Roger Casement*, p. 14.

p. 31: "Thinks, speaks well, most intelligent" from Joseph Conrad's Congo diary in *Heart of Darkness: An Authoritative Text, Backgrounds and Sources*, Robert Kimbrough, ed. (Englewood Cliffs, N.J.: Prentice Hall, 1960), p. 110.

p. 31: "I had been away from Ireland for years . . ." in René MacColl, *Roger Casement: A New Judgment* (New York: W. W. Norton & Co., 1957), p. 70.

p. 32: to "put a bridle on Spain . . ." in William Theobald Wolfe Tone, ed., *The Life of Theobald Wolfe Tone, written by himself and extracted from his journals* (London: Hunt & Clarke, 1828), pp. 32 and 43–5.

p. 32: "It was only because I was an Irishman . . ." in René MacColl, *Roger Casement*, p. 63.

p. 33: Casement's report on the Congo, reprinted in Singleton-Gates and Girodias, *The Black Diaries*, pp. 96–190.

p. 33: "It used to take ten days to get . . ." in Casement's Congo Report, reprinted in *The Black Diaries*, p. 112.

p. 34: "Brutal, savage, and barbaric torture . . ." in Elaine Scarry, *The Body in Pain:*

The Making and Unmaking of the World (New York and London: Oxford University Press, 1985), p. 38.

p. 34: "The task which the State agents . . ." excerpted from "Letter from the King-Sovereign of the Congo Free State to the State Agents, Brussels, 16th June, 1897," reprinted in Singleton-Gates and Girodias, *The Black Diaries*, p. 83.

p. 36: "September 30th . . ." in Singleton-Gates and Girodias, *The Black Diaries*, p. 251.

p. 36: "to relieve our feelings we began an elaborate . . . " in Reid, *The Lives of Roger Casement*, p. 110.

p. 37: "I said to this man that under the actual regime . . ." in Singleton-Gates and Girodias, *The Black Diaries*, p. 302.

p. 37: When T. F. Meagher . . . in Costigan, *A History of Modern Ireland*, p. 200.

p. 39: "When I landed in Ireland that morning . . ." in Reid, *The Lives of Roger Casement*, p. 351, Singleton-Gates and Girodias, *The Black Diaries*, pp. 413–14.

p. 41: sex with members of "the lowest orders" in Roger Sawyer, *Casement: The Flawed Hero* (London: Routledge & Kegan Paul, 1984): "In his far from celibate practices he satisfied his yearnings with partners who, in the main, came from the lowest orders of society" (p. 2) and "Another aspect of his sexual activities was that, with only one recorded identifiable exception, all his partners, regardless of nationality, were of the lowest social class. One can only estimate the effect of such intimate and frequent associations with those from a very different world on a man who was himself socially ambivalent; it threw him completely off balance . . ." (p. 145). Lack of landscape appreciation on pp. 47–8 of MacColl, *Roger Casement: A New Judgment*, where MacColl adds on p. 63, "My feeling is that Casement could rather easily have been steered away from the path of the treason which he finally chose, by the simple expedient of someone having been a bit nicer to him. All he really needed was to be flattered. . . ."

p. 43: "The apparition was always said to be of a kindly nature . . ." in Reid, *The Lives of Roger Casement*, p. 17, note D.

5 The Beggar's Rounds

p. 50: "History is a nightmare . . ." in James Joyce, *Ulysses* (Harmondsworth, England: Penguin, 1968), p. 13.

p. 50: "Amnesia is the true history of the new world" in Derek Walcott, "The Muse of History" in *The Post-Colonial Studies Reader*, Bill Ashcroft, Gareth Griffiths and Helen Tiffin, eds (London and New York: Routledge, 1995), p. 372.

p. 52: "And so we lost our history . . ." Sinéad O'Connor. Along with Salaman, *The History and Social Influence of the Potato*, the book *Seeds of Change: Five Plants that Changed Mankind* by Henry Hobhouse (New York: Harper and Row, 1986) documents some of the historical effect of potatoes and potato blight in Ireland.

p. 52: Kerby A. Miller, *Emigrants and Exiles: Ireland and the Irish Exodus to North America* (New York and Oxford: Oxford University Press, 1985), in the sections "Change: Ireland before the Great Famine" and "Continuity: The Culture of Exile."

p. 53: Joseph Lee, in *Irish Values & Attitudes: The Irish Report of the European Value Systems Study*, Michael Fogarty, Liam Rian, and Joseph Lee, eds (Dublin: Dominican Publications, 1984), p. 112.

p. 54: "Most countries send out oil, iron . . ." John F. Kennedy in Fintan O'Toole, *Black Hole, Green Card: The Disappearance of Ireland* (Dublin: New Island Books, 1994), p. 98.

p. 56: "It is obvious that Ireland's misfortune is . . ." in Engels, *History of Ireland*, in *Ireland and the Irish Question: A Collection of Writings by Karl Marx and Friedrich Engels* (Moscow: Progress Publishers, 1971), p. 174.

6 Anchor in the Road

p. 59: "Man is naturally a nomad . . ." J. M. Synge in Alan Price, ed., *J. M. Synge Collected Works: Prose* (London: Oxford University Press, 1966), pp. 195–6. See also J. M. Synge, *The Aran Isles and Other Writings* (New York: Vintage Books, 1962), especially the essays "The Vagrants of Wicklow," "In Wicklow," and "On the Road."

p. 60: "she saw a bird coming to her . . ." and subsequent quotes from "The Destruction of Da Derga's Hostel" in *Early Irish Myths and Sagas*, Jeffrey Gantz, trans. (Harmondsworth, England: Penguin, 1991), p. 64.

p. 62: "It blew a heavy gale . . ." in *The Life of Theobald Wolfe Tone*, p. 74.

p. 62: St Brendan's encounters from Geoffrey Ashe, *Land to the West: St. Brendan's Voyage to America* (New York: Viking Press, 1962), and D. P. Conyngham, *Lives of the Irish Saints* (n.p.: P. J. Kennedy and Sons, n.d.).

p. 68: "with a terrifying exactitude" in *Alexis de Tocqueville's Journey in Ireland July–August 1835*, Emmet Larkin, trans. and ed. (Washington, D.C.: Catholic University of America Press, 1990), pp. 91–2: "I found myself this morning [1 August 1835] on top of the coach beside an old Catholic . . . He went on from there to tell me what had been the fate of a great many families and a multitude of estates, passing from the time of Cromwell to that of William III with a terrifying exactitude and local memory. Whatever one does, the memory of the great persecutions is not forgotten. And when one sows injustice, he sooner or later reaps its fruits."

7 Wandering Rocks

p. 71: Re Drake and the potato: see Salaman, *The History and Social Influence of the Potato*, p. 147 and following. This is the classic work on potatoes and their transit from Peru to Ireland.

p. 72: "about roses growing out of . . ." Bob Dylan in Greil Marcus, *Dead Elvis* (New York: Doubleday, 1991), pp. 116–17. Some of the following statements about American musical history are also drawn from conversation with Marcus, with thanks.

p. 73: Re Timothy Murphy, see Hubert Howe Bancroft, *Register of Pioneer Inhabitants of California 1542–1848* (Los Angeles: Dawson's Bookshop, 1964), and volumes XX–XXII of *The Works of Hubert Howe Bancroft, History of California, Vols. 3–5* (Santa Barbara, Calif.: Wallace Hebberd, 1969, facsimile of the first editions) throughout which brief references to Murphy are scattered.

p. 73: For the Californios . . . the 1830s were something of a Golden Age: see Leonard Pitt, *The Decline of the Californios* (Berkeley: University of California Press, 1966); Alfred Robinson, *Life in California during a Residence of Several Years in that Territory. . . .* (Santa Barbara, Calif. and Salt Lake City, Utah: Peregrine Publishers, Inc.,

1970); and Neal Harlow, *California Conquered: War and Peace on the Pacific, 1846–1850* (Berkeley: University of California Press, 1982).

p. 74: the infamous Zimmermann Telegram: see Barbara Tuchman, *The Zimmermann Telegram* (New York: Viking Press, 1958).

p. 77: "The pear is near ripe for falling" in Harlow, *California Conquered,* p. 103.

p. 77: On Murphy's about-face, see Alan Rosenus, *General M. G. Vallejo and the Advent of the Americans* (Albuquerque: University of New Mexico Press, 1985), p. 167.

p. 77: Jerry Garcia's Olompali vision: Bill Barich, "The Last Transcendental Trip," *The New Yorker,* XXX October 11, 1993, p. 101.

8 Articles of Faith

p. 83: "above 150,000 Irish acres in Kerry" in Arthur Young, *Arthur Young's Tour in Ireland,* Arthur Wollaston Hutton, ed. (London: George Bell & Sons, 1892), vol. 1, p. 344.

p. 85: "The builders penetrated inland . . ." in Aubrey Burl, *The Stone Circles of the British Isles* (New Haven, Conn.: Yale University Press, 1976), p. 224.

9 A Pound of Feathers

p. 87: "There is something magnificently wild . . ." in Young, *Arthur Young's Tour in Ireland,* vol. 1, p. 348.

p. 88: "he leapt the wall . . .," Horace Walpole in "The History of the Modern Taste in Gardening" in John Dixon Hunt and Peter Willis, eds, *The Genius of the Place: The English Landscape Garden 1620–1820* (Cambridge, Mass.: MIT Press, 1988), p. 313.

p. 91: "a mobile army of metaphors . . ." in "On Truth and Falsity in Their Ultramoral Sense," *Collected Works of Friedrich Nietzsche,* vol. 2, Maximilian A. Mugge, trans. (New York: Russell and Russell, 1964), p. 190.

p. 92: "They are the oldest living things . . ." in Man Ray, *Self Portrait* (New York: McGraw Hill, n.d.), p. 356.

p. 92: ". . . when a European conceives of confronting nature . . ." in Joseph Brodsky, *New Yorker* essay on Robert Frost, 26 September 1994, p. 70.

p. 93: "When it comes to belief in 'the soul' . . .," Ryan in Fogarty, Rian, and Lee, *Irish Values & Attitudes,* p. 99: "By any standards, Ireland is still a preeminently religious country. More people attend church once a week than in any other country in the world. Asked how important God was in their lives, the Irish were far ahead of any nation in Europe. When it comes to belief in 'the soul,' in 'life after death,' in heaven, and in prayer, the Irish are so far ahead of the rest of the western world that any comparisons are totally irrelevant."

p. 93: John O'Donohue's *Stone as the Tabernacle of Memory* was published as a small book (without a publisher, but printed by Clodoiri Lurgan, Inverin, Co. Galway) in 1994.

10 And a Pound of Lead

p. 95: Re the history of Irish forests and deforestation: sources include Frank Mitchell, *The Irish Landscape* (London: Collins, 1976); F. H. A. Aalen, *Man and the Landscape in*

Ireland (London: Academic Press, 1978); Eoin Neeson, *A History of Irish Forestry* (Dublin: The Lilliput Press, 1993); Susan Powers Bratton, "Oaks, Wolves and Love: Celtic Monks and Northern Forests," *Journal of Forest History,* January 1989; "The Oakwoods of Killarney," brochure by the Office of Public Works, Dublin; Alan Craig, "Woodland Conservation in Killarney National Park and Elsewhere in Ireland," National Parks and Wildlife Service, Dublin, 1992 (unpublished).

p. 95: "no less cautions . . . ," Fynes Moryson in Nicholas Canny, *Kingdom and Colony: Ireland in the Atlantic World 1560–1800* (Baltimore, Md.: Johns Hopkins University Press, 1988), p. 2 . See also *The Westward Enterprise: English Activities in Ireland, the Atlantic, and America 1480–1650,* K. R. Andrews, Nicholas Canny, and P. E. H. Hair, eds (Liverpool and Detroit, Mich.: Wayne State Press, 1978).

p. 95: "It was ominous for both . . ." in Canny, *Kingdom and Colony,* p. 35.

p. 95: "Fraught with all vice . . ." in David Beers Quinn, *The Elizabethans and the Irish* (Ithaca, N.Y.: Cornell University Press, 1966), pp. 135–6: "Gervase Markham had just returned from Irish service in 1600 when he wove a long episode about the Irish kern into his rambling poem, 'The New Metamorphosis.'"

p. 95: See Edmund Spenser's *A View on the Present State of Ireland,* reprinted in James P. Myers, Jr, ed., *Elizabethan Ireland: A Selection of Writings by Elizabethan Writers on Ireland* (Hamden, Conn.: Archon Books, 1983); and Patricia Coughlan, ed., *Spenser and Ireland: An Interdisciplinary Perspective* (Cork: Cork University Press, 1989).

p. 95: George Percy in Quinn, *The Elizabethans and the Irish,* p. 23, along with many other such analogies.

p. 95: "I defy you, my dear cousin . . ." in Tocqueville, *Alexis de Tocqueville's Journey in Ireland,* p. 7.

p. 96: "In a few more years, a Celtic . . ." in *The Times,* quoted in Miller, *Emigrants and Exiles,* p. 307.

p. 96: See James Mooney, *The Ghost Dance Religion and the Sioux Outbreak of 1890,* republished with an introduction by Bernard Fontana (Glorieta, N. Mex.: Rio Grande Press, 1973).

p. 96: "starvation and squalor caused an outbreak . . ." in Sawyer, *Casement: The Flawed Hero,* p. 92.

p. 96: Ward Churchill in *Indians Are Us? Culture and Genocide in Native North America* (Monroe, Maine: Common Courage Press, 1994), pp. 234, 310 and 342, note 45.

p. 97: The Irish . . . had a long history of comparing themselves to the Jews. . . . See Canny, *Ireland in the Atlantic World,* p. 111: ". . . many priests had . . . sought to 'comfort their flocks partly by prophesies of their restoration to their ancient estates and liberties . . . by way of God's promise to restore the Jews and the kingdom of Israel.'" The same analogy is used in Joyce, *Ulysses,* p. 143. See also Pat Feely, "Aspects of the 1904 Pogrom" in *Old Limerick Journal* 11, Winter 1992.

p. 97: "*Low-browed* and *savage* . . ." in David Roediger, *The Wages of Whiteness: Race and the Making of the American Working Class* (London: Verso, 1991), p. 133.

p. 97: "The people are thus inclined . . ." in Edmund Campion, *A History of Ireland,* in Myers, ed., *Elizabethan Ireland,* p. 24.

p. 98: "The royal roads were cow paths . . ." in Seamus Heaney, *Station Island* (London: Faber & Faber, 1984), p. 101.

p. 98: "I believe that seasonal nomadism . . ." in E. Estyn Evans, *Irish Folk Ways* (London: Routledge, 1988, first printed in 1957), p. 27. See also Nicholas Canny, "Early Modern Ireland," in *The Oxford Illustrated History of Ireland* (Oxford and New York: Oxford University Press, 1989), pp. 108–9: "Pastoralism continued to dominate. . . . Seminomadic pastoralism on unenclosed countryside was particularly suited to unsettled political conditions."

p. 98: ". . . to keep their cattle and to live themselves . . ." in Spenser, *A View on the Present State of Ireland*, in Myers, ed., *Elizabethan Ireland*, p. 79.

p. 99: "The hereditary status of the learned . . ." in Quinn, *The Elizabethans and the Irish*, p. 11

p. 99: "The reasons that entitled a woman . . ." in Engels, *Origin of the Family, Private Property and the State*, excerpted in *Ireland and the Irish Question: A Collection of Writings by Karl Marx and Friedrich Engels*, p. 339.

p. 101: "Save them, says the citizen . . ." in Joyce, *Ulysses*, p. 325.

p. 101: "whosoever could take a rhymer . . ." Thomas Churchyard in Quinn, *The Elizabethans and the Irish*, pp. 126–7.

p. 101: "rhymers swore to rhyme these gentlemen. . . ." in Quinn, *The Elizabethans and the Irish*, pp. 126–7.

p. 102: "He was someway gifted . . ." in Lady Gregory, *Poets and Dreamers: Studies and Translations from the Irish, including Nine Plays by Douglas Hyde* (Gerrard's Cross, England: Colin Smythe, 1974), p. 16.

p. 102: woods "of matchless height" . . . in Costigan, *A History of Modern Ireland*, p. 58; see also Fabricant, *Swift's Landscape*, pp. 90–93.

p. 103: Re the arguments over the Golden Age, see, for example, Merlin Stone, *When God Was a Woman* (San Diego, Calif.: Harcourt Brace Jovanovich, 1976); Riane Tennenhaus Eisler, *The Chalice and the Blade* (New York: Harper and Row, 1987); Robert Bly, *Iron John: A Book about Men* (Reading, Mass.: Addison-Wesley, 1990); Sam Keen, *Fire in the Belly: On Being a Man* (New York: Bantam Books, 1991); and Ward Churchill's fine response to the men's movement, the title essay in his *Indians Are Us?* Re Arcadia: Seamus Heaney, "In the Country of Convention: English Pastoral Verse" and "The God in the Tree: Early Irish Nature Poetry" in *Preoccupations* (London: Faber & Faber, 1980), the chapter "Ireland as Arcadia" in Quinn, *The Elizabethans and the Irish*, the chapter "Antipastoral Vision and Antipastoral Reality" in Fabricant, *Swift's Landscape*; and Declan Kiberd, "Inventing Irelands" in *Crane Bag*, vol. 8, no. 1, 1984.

p. 104: "It would seem probable . . ." in Terry Eagleton, *Heathcliff and the Great Hunger* (London: Verso, 1995), p. 6.

p. 104: Swift's "A Pastoral Dialogue" in Davis, ed., *Swift: Poetical Works*, pp. 393–5.

p. 105: Swift's "Verses occasioned by the sudden drying up of St. Patrick's Well . . . ," in Davis, ed., *Swift: Poetical Works*, p. 385.

p. 105: "Ill fares the land" p. 1653 and "Sweet Auburn! parent of" p. 1654 of M.H. Abrams, general editor, *The Norton Anthology of English Literature*, vol. 1 (New York: W.W. Norton & Co., 1962).

pp. 105–6: "the Livelyhood of a Hundred People," Swift in Fabricant, *Swift's Landscape*, p. 83.

p. 106: "A Pastoral Ballad by John Bull" in John Montague, *The Book of Irish Verse:*

An Anthology of Poetry from the Sixth Century to the Present (New York: Macmillan, 1974), p. 191.

p. 106: Patrick Kavanagh, *The Great Hunger* (London: MacGibbon and Kee, 1966), pp. 13, 20.

11 The Circulation of the Blood

p. 110: Joyce once remarked . . . in Richard Ellman, *James Joyce* (New York: Oxford University Press, 1959), p. 230.

p. 112: Wallace Stevens, "The Irish Cliffs of Moher" in Wallace Stevens, *The Collected Poems* (New York: Vintage Books, 1982), p. 503.

p. 114: For the etymology of the word *race*, see Michael Burleigh and Wolfgang Wippermann, *The Racial State: Germany 1933–1945* (Cambridge: Cambridge University Press, 1991), p. 23: "The word *Rasse* (race) is thought to derive from the Arabic *ras* (meaning, 'beginning', 'origin', 'head'). It entered the German language in the seventeenth century, as a loan word from English and French. . . ."

p. 114: "it is worth adding . . ." in Eagleton, *Heathcliff and the Great Hunger*, p. 279.

p. 115: Re the two bodies of the king: see Laurie A. Finke, "Spenser for Hire: Arthurian History as Cultural Capital" in *Culture and the King: The Social Implications of the Arthurian Legend*, Martin B. Shichtman and James P. Carley, eds (Albany: State University of New York Press, 1994).

p. 115: "It is one thing to sing . . ." Rainer Maria Rilke, "Third Duino Elegy," in *The Selected Poetry of Rainer Maria Rilke*, Stephen Mitchell, ed. and trans. (New York: Vintage, 1989), p. 163.

p. 115: twenty zealots who took blood seriously . . . in Costigan, *A History of Modern Ireland*, p. 290.

p. 115: Ethnic Germans who settled . . . in Michael Ignatieff, *Blood and Belonging: A Journey into the New Nationalism* (New York: Farrar, Straus, & Giroux, 1994).

p. 115: Rilke, "Third Duino Elegy," in *The Selected Poetry of Rainer Maria Rilke*, p. 165.

p. 116: "How many diseases have their origin . . ." and "made no distinctions . . ." in Michael Burleigh and Wolfgang Wippermann, *The Racial State*, pp. 40 and 107.

p. 116: "a skin, mediating the mineral . . ." in Paul Shepard, *Thinking Animals: Animals and the Development of Human Intelligence* (New York: Viking Press, 1978), p. 4.

p. 117: "If, to use tempting older . . ." in David Roediger, *The Wages of Whiteness*, p. 8.

p. 117: "Trees have roots. . . ." in Simon Schama, *Landscape and Memory* (New York: Alfred A. Knopf, 1995), p. 29.

12 Rock Collecting

p. 121: "I have my ancestors' pale blue eyes . . ." in Arthur Rimbaud, *Une Saison en Enfer/ A Season in Hell*, Louise Varèse, trans. (New York: New Directions, 1945), p. 7: "J'ai de mes ancêtres gaulois l'oeil bleu blanc, la cervelle étroit. . . . Mais je ne beurre pas ma chevelure."

13 The War between the Birds and Trees

p. 128: Lynn White, "The Historical Roots of Our Ecological Crisis," in Paul Shepard and Daniel McKinley, eds, *The Subversive Science: Essays towards an Ecology of Man* (Boston: Houghton Mifflin, 1969), pp. 341–50. See also Paul Shepard, *Nature and Madness* (San Francisco: Sierra Club Books, 1982).

p. 128: The Greening of the Church is also the title of a book by Sean McDonagh (London: G. Chapman/Maryknoll; New York: Orbis Books, 1990).

p. 132: "Has anyone ever considered . . ." in George Santayana, "The Philosophy of Travel" in *Altogether Elsewhere: Writers on Exile*, Marc Robinson, ed. (New York: Harcourt Brace, 1994), p. 41.

14 Wild Goose Chase

p. 133: The story of the Children of Lir is in *Old Celtic Romances Translated from the Gaelic by P. W. Joyce,* second edition, revised and enlarged (London: D. Nutt, 1894), pp. 2–36.

p. 133: "The Irish tradition that . . ." Artelia Court, *Puck of the Droms* (Berkeley: University of California Press, 1985), pp. 236–7, note 11.

p. 133: The Frenzy of Sweeney exists in its entirety in one edition other than Seamus Heaney's "free translation": the 1913 Irish Texts Society bilingual edition with translation, introduction, and notes by J. G. O'Keeffe. Unless otherwise noted, quotes are from his translation.

p. 134: "a figure of the artist . . ." in Seamus Heaney, *Sweeney Astray: A Version from the Irish* (New York: Farrar, Straus, & Giroux, 1983), second page of unpaginated introduction.

p. 134: "Cowardly men run wild . . ." in O'Keeffe, *Buile Suibhne* (*The Frenzy of Sweeney*) (London: Irish Texts Society, 1913), p. xxxiv, note 2.

p. 135: "Dreams of flying or floating . . ." in Sigmund Freud, *The Interpretation of Dreams*, James Strachey, ed. and trans. (New York: Avon Books, 1965), p. 429.

p. 136: For the jail letters and drawings of Countess Markievicz, see Ann Haverty, *Constance Markievicz: An Independent Life* (London: Pandora, 1988).

p. 136: "When the soul of a man . . ." in Joyce, *Portrait of the Artist* (Ware, Hertfordshire: Wordsworth Classics, 1992), p. 203.

p. 137: "If traditional myths . . ." in Fintan O'Toole, *A Mass for Jesse James: A Journey through 1980's Ireland* (Dublin: Raven Arts Press, 1990), p. 13.

p. 137: "Who *is* Sweeney . . ." in Nevill Coghill, "Sweeney Agonistes (An Anecdote or Two)," *Critical Essays on T. S. Eliot: The Sweeney Motif*, Kinley E. Roby, ed. (Boston: G.K. Hall, 1985), p. 119.

p. 137: Herbert Knust, "Sweeney among the Birds and Brutes," *Critical Essays on T.S. Eliot*, pp. 169 ff.

p. 138: "Apeneck Sweeney . . ." and other poems in T. S. Eliot, *Selected Poems* (New York: Harcourt, Brace & World, 1964).

p. 138: "swine" and "ravenous bitch": see Nancy D. Hargrove, "The Symbolism of Sweeney in the Work of T.S. Eliot," *Critical Essays on T. S. Eliot*, p. 150: "First the name 'Sweeney' not only has a sound which is common, prosaic, unmusical, perhaps even vulgar, but also it evokes the word 'swine' with its connotations of bestial and gross physicality, ugliness, dirt, and stupidity;" and George Whiteside, "A Freudian Dream Analysis of 'Sweeney among the Nightingales,' " p. 64: "it sounds like 'Raven-a-bitch': a bitch dog with ravenous hunger." In *Critical Essays on T.S. Eliot.* Since I wrote this chapter, Anthony Julius's *T. S. Eliot, Anti-Semitism and Literary Form* (New York: Cambridge University Press, 1995) has done much to confirm and explore Eliot's anti-Semitism.

p. 138: "in manipulating a continuous parallel . . ." from Eliot, "*Ulysses*, Order and Myth" in *Dial* lxxv, November 1923, reprinted in *James Joyce: The Critical Heritage*, vol. 1 (New York: Barnes and Noble, 1970), p. 270.

p. 138: For *The Waste Land* as a parody of *Ulysses*, see the note on p. 160 of William York Tindall's *A Reader's Guide to James Joyce* (New York: Farrar, Straus, & Giroux, 1959).

p. 138–9: "So complicated in his thought . . ." in Ellman, *James Joyce*, p. 165.

p. 139: "some thin little bird . . ." Harold Nicolson, in Brenda Maddox, *Nora: The Real Life of Molly Bloom* (Boston: Houghton Mifflin, 1988), p. 271.

p. 139: "There are sixteen geese . . ." in Maddox, *Nora*, p. 380.

p. 139: "seemed like one whom magic . . ." in Joyce, *Portrait*, p. 171.

p. 139: "What birds were they?" and following in Joyce, *Portrait*, pp. 224–6.

p. 139: Re the Wandering Jew, see George K. Anderson, *The Legend of the Wandering Jew* (Providence, R.I.: Brown University Press, 1965); Galit Hasan-Rokem and Alan Dundes, eds, *The Wandering Jew: Essays in the Interpretation of a Christian Legend* (Bloomington: Indiana University, 1986); and Ira B. Nadel, *Joyce and the Jews: Culture and Texts* (Iowa City: University of Iowa Press, 1989).

p. 140: "a hawk-like man . . ." in Joyce, *Portrait*, p. 169.

p. 141: "Forget wings . . ." in Neil Jordan, "The Dream of a Beast," *The Neil Jordan Reader* (New York: Vintage, 1993), p. 158.

p. 141: Heaney, *Sweeney Astray*, last page of unpaginated intro.

15 Grace

p. 144: German and English aristocrats . . . had to hire hermit impersonators. . . . See Christopher Thacker, *The History of Gardens* (Berkeley: University of California Press, 1979).

p. 145: "And Patrick went forth . . ." in Newport J. D. White, *Saint Patrick: His Writings and Life* (London and New York: Macmillan Co., 1920), p. 8.

p. 146: The only biography of substance on Grace O'Malley seems to be Anne Chambers's *Granuaile: The Life and Times of Grace O'Malley* (Dublin: Wolfhound Press, 1988).

p. 147: "She is now most vividly . . ." in Chambers, *Granuaile*, p. 178.

p. 148: "A chieftain of a neighboring clan . . ." in Chambers, *Granuaile*, p. 76.

16 Travellers

Among the works consulted but not cited were: *Traveller Ways, Traveller Words* (Dublin); *Do You Know Us at All?* by the Parish of the Travelling People, Dublin; *Anti-Racist Law and the Travellers* by the Irish Traveller Movement (Dublin, 1993); J. P. Liégeois, *Gypsies and Travellers* (Strasbourg: Council on Cultural Co-Operation, 1987); Judith Okely, *The Traveller Gypsies* (New York: Cambridge University Press, 1983); Sharon Gmelch, *Tinkers and Travellers: Ireland's Nomads*, with photographs by Pat Langan and George Gmelch (Montreal: McGill-Queen's University Press, 1976); George Gmelch, *To Shorten the Road: Essays and Biographies*, with folktales edited by Ben Kroup (Dublin: The O'Brien Press, 1978); Donald Kenrick and Grattan Puxon, *The Destiny of Europe's Gypsies* (New York: Basic Books, 1973); Lady Gregory, "The Wandering Tribe," *Poets and Dreamers*; the essays on wanderers in J. M. Synge, *The Aran Islands and Other Works*; Synge's *The Tinker's Wedding;* and the *Journal of the Gypsy Lore Society.*

p. 152: For the bishop of Galway's offer, see *The Irish Times*, 8 February 1994.

p. 153: Nan Joyce, *Traveller: an Autobiography*, Anna Farmar, ed. (Dublin: Gill and Macmillan Ltd., 1985).

p. 155: "Their very existence constituted dissidence," Jean-Pierre Liégeois, *Gypsies: An Illustrated History* (London: Al Saqi Books, 1986), p. 104.

p. 156: For information on Travellers in the United States, see the work of Jared Harper, cited in May McCann, Seamas O'Siochan, Joseph Ruane, *Irish Travellers: Culture and Ethnicity* (Belfast: Institute of Irish Studies, Queen's University of Belfast, 1994).

p. 157: For cant and gammon, see Sinéad Ní Shuinear, "Irish Travellers, Ethnicity and the Origins Question," in *Irish Travellers: Culture and Ethnicity*, p. 135: "the Shelta word cuinne 'a priest' is an old word for druid."

p. 157: "In their deep religious feeling . . ." in the National Federation of the Irish Travelling People report cited in *The Irish Times*, 2 February 1994.

p. 157: "seemed so premodern . . ." in Kerby Miller, *Emigrants and Exiles*, p. 107.

p. 158: "antiquated traditions . . ." in Court, *Puck of the Droms*, p. 1.

p. 158: "Some nomadic peoples . . ." in Ní Shuinear, "Irish Travellers, Ethnicity and the Origins Question," *Irish Travellers: Culture and Ethnicity*, p. 60.

p. 158: Gilles Deleuze and Felix Guattari on rhizomes in *On the Line*, John Johnson, trans. (New York: Semiotexte, 1983). Another of their Semiotexte books, *Nomadology: The War Machine* (1986), though opaque and trendy at times, is germane.

p. 159: "for Travellers, the physical fact . . . " in Michael McDonogh, "Nomadism in Irish Travellers' Identity," in *Irish Travellers: Culture and Ethnicity*, pp. 95–6.

p. 160: "The life of insecurity . . ." in Freya Stark, *The Journey's Echo* (London: J. Murray, 1963), p. 168.

p. 161: "For a woman a house is . . ." in *Irish Tinkers*, photographed and compiled by Janine Wiedel, with a foreword and transcripts by Martina O'Fearadhaigh (New York: St. Martin's Press, 1978), p. 61.